THE INHERI[barcode: I0069215] OF TRAITS

From Genetics to Heredity

Edited By: Kashaf Noreen

Written By:
Austin Mardon
Natalie Wong
Michael Tang
Hanna Redda
Nataliya Raza
Hareem Bilal
Vivek Kannan
Natalie Jean-Marie
Mical Habtemikael
Mackenzie Schuler
Mya Elisabeth George
Navneet Kang

II

THE INHERITANCE OF TRAITS

From Genetics to Heredity

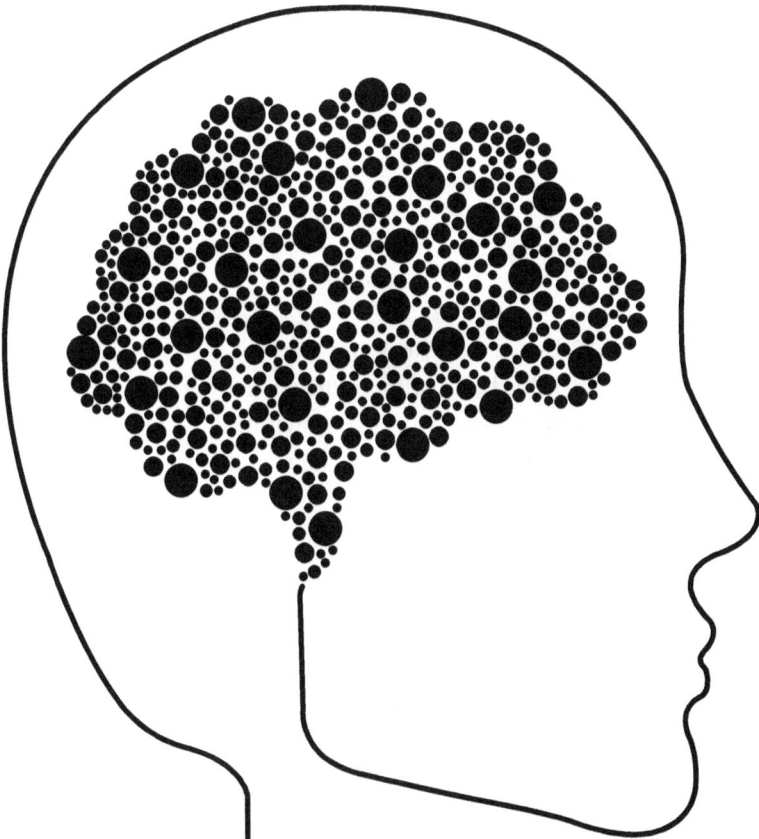

Austin Mardon, Kashaf Noreen, Natalie Wong, Michael Tang, Hanna Redda, Nataliya Raza, Hareem Bilal, Vivek Kannan, Natalie Jean-Marie, Mical Habtemikael, Mackenzie Schuler, Mya Elisabeth George, Navneet Kang

III

Typeset and Cover Design by Summer Shields

ISBN 978-1-77369-260-9

Golden Meteorite Press
103 11919 82 St NW
Edmonton, AB T5B 2W3
www.goldenmeteoritepress.com

Contents

What is Heredity?

Written By: Natalie Wong

What is Heredity?

The concept of heredity is involved in many parts of our lives, with it being responsible for the makeup of our very being and the traits that all living organisms exhibit. So what is heredity? Heredity, also referred to as biological inheritance, is the passing down of genetic information from parent organisms to their offspring in order to specify traits and functions.[1] This process of acquiring the genetic material of parental organisms can be achieved through asexual or sexual reproduction, and is not limited to humans. Heredity and reproduction are fundamentals to how life is defined, being distinguished from other natural processes that do not possess a link between parent and organism.[2] For example, the formation of ocean waves does not require a parent organism to exist. Hereditary information specificies many aspects of our lives, ranging from chemical processes within our bodies to anatomical features such as our eye colours.[3] The study of heredity is also recognized as the

1 Alberts, 2015
2 Alberts, 2015
3 Alberts, 2015

1

field of genetics, and these two terms are often used interchangeably. In addition to observing how traits are passed from parent to offspring, the study of genetics also examines how one's environment and experiences influence development, appearance, and behaviour. The field of genetics has significant applications from medicine to biotechnology, with notable developments such as CRISPR and gene therapy.

What is Inherited and What is Not: Genotype and Phenotype

Traits that are inherited and passed on to the next generation are known as heritable traits, and are controlled by a single gene or a set of genes. All heritable traits are governed by our genotype, which is the complete set of genetic material, or genes, that an organism possesses.[4] Examples of inheritable traits include eye colour, and albinism. We can inherit a certain eye colour, or the condition of albinism depending on whether our parents have passed on the genes for these traits or not. If you're interested in determining your genotype, the process of genotyping ca help you describe your entire genetic makeup.[5]

However, not all traits that we are able to observe physically in the form of the structure or behaviour of the organism are inheritable. Phenotype refers to the physical properties of an organism that are observable, such as behaviour, appearance, and development.[6] Phenotype is determined by one's genotype as well as the interactions these genes have with the environment. Therefore, various aspects of one's phenotype are not inherited due to the influence of environmental factors that may lead to a phenotype that differs from its genotype.[7] For example, height is a phenotype as it is dictated by genetic material passed on from our parents that may predispose us to a certain height, but can also be influenced by environmental factors such as the availability of nutrition. The relationship between genotype and phenotype is demonstrated in identical twins. Although identical twins possess identical genotypes that were passed on from their parents, the influence of environmental factors during development led to the expression of phenotypes that are not

---

4 Genotype , n.d.

5 Genotype , n.d.

6 Phenotype / Phenotypes , n.d.

7 Phenotype / Phenotypes , n.d.

identical.[8]

How is Hereditary Information Stored: Genes, Alleles, and Chromosomes

Hereditary information, or the genetic instructions to produce or maintain a living organism, is stored in all living cells in the form of genes.[9] A gene is a segment of DNA that encodes an mRNA strand that is subsequently translated into a polypeptide chain, which will be folded to form a functional protein.[10] The protein will then carry out the instructions as dictated by their genetic material. Examples of the wide range of instructions from our genetic material include biochemical processes within the body, our eye colour, and our height. Within the nucleus of animal and plant cells, chromosomes, which are structures composed of protein and coiled DNA, act as a key player in heredity.[11] Particularly, the role of DNA is central to the passing on of genetic information.

There can be more than one variation of the same gene at the same location on a chromosome. Alleles refers to a different, or variant form of a gene that an organism possesses. Some genes have different forms located at the same genetic locus, which is the position on a chromosome.[12] Humans have a pair of alleles at each genetic locus, with one inherited from each parent.[13] These characteristics classify humans as diploid organisms.

Whether the pair of alleles we are examining at a particular locus are identical or not is central to how different variations of a trait can be passed onto the next generation. If the genotype contains two of the same alleles at a locus, it is considered homozygous. If the two alleles are not identical, they are considered homozygous.[14] In addition to the contribution of alleles to one's genotype, it is also involved in one's phenotype.

---

8 Phenotype / Phenotypes , n.d.
9 Alberts, 2015
10 Pearson, 2006
11 Cooper, 2000
12 Allele, n.d.
13 Allele, n.d.
14 Allele, n.d.

Our current understanding of genes comes from discoveries of various scientific disciplines. Within the field of classical genetics, genes were an abstract unit that conferred inheritance of traits from the parent organism to offspring.[15] The field of biochemistry associated one enzyme or protein for each gene, while molecular biology led to the current definition of a gene as a DNA segment.

The Central Dogma

The central dogma is a foundational principle in molecular biology that outlines the transfer of genetic information in cells from DNA, to RNA, and then to protein.[16] Although this principle is considered universal, there is variation between organisms in how genetic information flows. In eukaryotic cells, RNA transcripts are processed within the nucleus before they can travel to the cytoplasm for protein translation.[17] These steps of processing include RNA splicing, which can change the instructions of an RNA molecule.[18] The resulting implication is crucial for understanding how eukaryotic cells read their genomes.

All About DNA

As mentioned previously, DNA is central in the concept of hereditary as it consists of the genetic material of the cell. Also known as deoxyribonucleic acid, DNA is a double helix consisting of two long polypeptide chains held together by hydrogen bonds.[19] It is composed of four types of nucleotide subunits that has a deoxyribose sugar attached to a single phosphate group, with one of the four types of nitrogenous bases: A (adenine), T(thymine), C (cytosine), and G (guanine).[20] In essence, the linear sequence of nucleotides within DNA is how genetic information is carried in a chemical form and this genetic information dictates the formation of proteins and RNA molecules that an organism will need to synthesize in order to create and maintain life.

---

15 Pearson, 2006
16 Alberts, 2015
17 Alberts, 2015
18 Alberts, 2015
19 Alberts, 2015
20 Alberts, 2015

The unique double stranded structure of DNA provides a mechanism for how genetic material is copied then transmitted to the next generation in hereditary.[21] Each DNA strand has a nucleotide sequence that is complementary to the sequence of the partner strand.[22] For nucleotide base pairs to be complementary, A nucleotides must always pair with T nucleotides, while C nucleotides must bond to G nucleotides. In the process of DNA replication, each DNA strand can act as a template for the synthesis of a new identical complementary DNA strand.[23] This allows for an accurate replication of the genetic information in DNA to pass it to the next generation.

Discovery of the Role of DNA in Hereditary

Initial evidence on the role of DNA was discovered through studies on the pneumonia bacteria, where it was discovered that there was a substance in heat-killed encapsulated bacteria which induced genetic transformation of nonencapsulated bacteria to their encapsulated counterparts.[24] Another key study that led to the acceptance of DNA as genetic material was conducted by Oswald Avery, Colin MacLeod, and Maclyn McCarty in 1944. They established that DNA was the transforming principle through selective enzymatic digestion of different cell components.[25] Later experiments with bacterial viruses solidified the role of DNA through the discovery that the component that must enter the host cell for viral replication is viral DNA, and not other components such as proteins.[26] The likes of these experiments confirmed that the genetic material of the cell is indeed DNA.

The Role of RNA in Hereditary

According to the central dogma, RNA acts as an intermediate during the expression of hereditary information to ensure proteins are ultimately made to carry out instructions as dictated by the genome.[27] However,

---

21 Alberts, 2015
22 Alberts, 2015
23 Alberts, 2015
24 Cooper, 2000
25 Cooper, 2000
26 Cooper, 2000
27 Pearson, 2006

for many genes, RNA is the final coded product instead of proteins.[28] RNA, similar to proteins, is able to fold into three dimensional structures, provide a structural function and catalyze biochemical reactions within the cell, as well as acting to regulate gene expression.[29] Additionally, it actively regulates processes in the cell, and may even act as the genetic material to transfer information across generations.[30]

A study from 2005 suggested that plants can occasionally rewrite their DNA based on messages in RNA molecules inherited from past generations.[31] With suggestions that a similar phenomenon may also occur in mice, and by extension possibly in other mammals, this poses significant implications in the field of evolutionary genetics.[32] Although mechanisms of heredity may seem established, further research can reveal novel discoveries which may completely change the course of how we see heredity.

Transfer of Genes Within a Species: Sexual Reproduction and Horizontal Gene Transfer

The transfer of genetic information within a species can be achieved through two modes: horizontal gene transfer and sexual reproduction. Horizontal gene transfer, which is the process of transferring genetic information between two different cell lineages, results in organisms that are more closely related to either set of relatives with respect to a variety of genes.[33] Sexual reproduction is responsible for the vertical transfer of genetic material, where genetic information is transmitted from parents to offspring.[34] Additionally, genetic exchange occurs during sexual reproduction, allowing for genetic diversity. Sexual reproduction is a common form of reproduction, particularly in eukaryotes, but asexual reproduction also occurs and will be dsicussed next.[35]

---

28 Alberts, 2015
29 Alberts, 2015
30 Pearson, 2006
31 Pearson, 2006
32 Pearson, 2006
33 Alberts, 2015
34 Alberts, 2015
35 Alberts, 2015

Recombination During Meiosis

Meiosis is a type of cell division used to produce gametes, or germ cells (the sperm and the egg), from a single cell in organisms that undergo sexual reproduction.[36] Most plant and animal cells of higher organisms are diploid, meaning that they have two copies of each chromosome.[37] In contrast, germ cells, such as the sperm and the egg, are haploid. Haploid organisms only possess one copy of the chromosome. One member of each chromosome pair derived from the respective parents is transferred to each daughter cell in meiosis.[38] Through the union of the haploid sperm and egg in the process of fertilization, a diploid organism is created.[39] Genetic recombination occurs during meiosis, where members of chromosome pairs exchange genetic material to result in recombination between linked genes.[40] The frequency of genetic recombination depends on the distance between the linked genes, where a reduced distance will allow more frequent recombination.[41]

Asexual Reproduction

In asexual reproduction, the process of binary fission is used for cell division and the transfer of genetic material instead of meiosis or mitosis. Binary fission is the method in which simpler organisms, such as yeasts, conduct cell division.[42] It results in the production of new individuals which are genetically identical to the parent at all locations in the genome except at sites with somatic mutations.[43] In essence, asexual reproduction results in the transfer of genetic material that is identical to the parent, to the progeny, without the involvement of genetic recombination.

Transfer of Genes Between Different Species

Genes can be transferred between organisms of different species through

---

36 Cooper, 2000
37 Cooper, 2000
38 Cooper, 2000
39 Cooper, 2000
40 Cooper, 2000
41 Cooper, 2000
42 de Meeûs et al., 2007
43 de Meeûs et al., 2007

the mechanism of horizontal gene transfer. This process often occurs in prokaryotes and rarely in eukaryotes.[44] An example of horizontal gene transfer in prokaryotes involves viruses as a vector for gene transfer. Some genetic sequences within prokaryotes are derived from bacteriophages, a type of virus that infects bacteria. In the process of viral replication, often the infected host cells that are of the same or different species are killed.[45]

However, on the occasions that they are not, the viral DNA persists in the host cell for multiple generations without any effect on the host.[46] The viral DNA may persist as a plasmid, which is a separate DNA fragment within the cell, or as a DNA sequence embedded into the host cell's genome.[47] Additionally, viruses can promote genetic diversity by transferring DNA fragments of one host cell genome into another. In the field of medicine, the role of horizontal gene transfer is central to the spread of antibiotic resistance genes, where they are able to transfer between species and therefore provide the recipient bacteria with an advantage to survive even in the presence of antibiotics.[48] As a result, lethal and antibiotics resistant bacteria strains are able to survive and cause harm even in the presence of antibiotics that are otherwise able to eliminate the pathogens.

Plant Heredity

Although the genetics of plants are similar to that of animals in multiple ways, including the usage of DNA as genetic material to pass on traits, there are several key differences. The ability of plants to self fertilize adds a layer of complexity as it results in the loss of genetic diversity and potentially inbreeding depression, which refers to the reduced fitness of offspring that are inbred.[49] Another unique aspect of plants is their ability to undergo photosynthesis. Photosynthesis is the process by which plants and other organisms convert light energy from the sun into chem-

---

44 Alberts, 2015
45 Alberts, 2015
46 Alberts, 2015
47 Alberts, 2015
48 Alberts, 2015
49 Wright et al., 2013

ical energy through cellular respiration to be used by the plant as an energy source. This process is activated by chloroplasts, also known as plastids, that possess their own DNA (cpDNA).[50] With an additional reservoir of genetic material compared to animals, plants have increased additional genetic diversity and complexity.

Whenever plant genetics are mentioned, your mind may have gone straight to Gregor Mendel and his pea plant breeding experiments. Mendel, who deduced several classical genetic principles to explain how traits are passed down in 1865, is a major figure in the field of genetics.[51] In his studies of the inheritance of various plant traits, including seed colour, he was able to predict and interpret the inheritance patterns that were observed through assumptions that a trait is determined by a pair of genes.[52] Currently, experiments in plant genetics use different model organisms, such as Arabidopsis Thaliana, to study plant hereditary and development.

Arabidopsis is well suited for genetic analysis due to its small genome size, small size of the mature plant with short generation time and self fertilizing flowers.[53] In addition, the large amounts of progeny made per plants are perfect for mutant screens.[54] The applications of plant genetics have significant impacts on the economy as crops can be modified genetically to yield favourable characteristics, such as increased nutritional values, higher yields, resistance to harsh environmental conditions. An example includes the widely controversial subject of genetically modified organisms, particularly those slated for human consumption due to safety concerns.

Conclusion

All of life is dependent on the concept of hereditary. Without it, cells are unable to utilize the genetic instructions from the parental organism to carry out processes that are required for the creation of living organisms and their maintenance. However, genetics is not a stand-alone

---

50 de Vries & Archibald, 2018
51 Cooper, 2000
52 Cooper, 2000
53 Koornneef & Meinke, 2010
54 Koornneef & Meinke, 2010

field. With its relation to the theory of evolution, and the rise of various sub-disciplines such as population genetics that will be discussed in the following chapters, it is ever-evolving and can lead to significant applications in human health.

Over decades of research in genetics, we have mapped out the entire human genome,[55] found underlying genetic markers to various diseases,[56] and even utilized the concept of heredity in biotechnology to improve agricultural production. With the current trajectory of genetics research, the concept of heredity will definitely transform from a simple concept of how we pass genetic information to offspring into endless possibilities of understanding and improving the world that we live in.

---

55 The Human Genome Project, n.d.
56 Kim et al., 2014

The Relationship Between Genetics and Evolution

Written By: Michael Tang

What is Evolution, and What is its Relationship to Genetics?

Evolution refers to any physical and behavioural changes in organisms that occur over time, often directed or influenced by non-random forces (such as natural selection) and random forces (such as random mutations).[57] However, changes cannot be passed on through generations without some form of information which can code for these traits and can be passed on to subsequent generations. DNA, one of the fundamental biomolecules involved in genetics and heredity, checks both of these boxes. In other words, evolution refers to the inheritance of changes that are selected for or favoured by environmental factors or as the result of random chance or mutations. Genetics determines most if not all of an organism's physical traits and behaviours and provides the avenue for particular traits to be passed onto future generations based on how well they are suited for ensuring survival and reproductive success.

The preeminent and most widely accepted theory of evolution is Charles Darwin's theory of evolution by natural selection, which was first pro-

---

57 Than, 2018

posed in his book On the Origin of Species by Means of Natural Selection, which was published in 1859.[58] Darwin's theory of evolution spawned "Darwinism," a way of thinking about evolution that is based on principles such as natural selection. Darwinism has in turn both guided and inspired other evolutionary theorists in thinking about evolution and adding more concepts to this way of thinking.[59] The remainder of this chapter will focus on some of the major concepts within and related to Darwin's theory of evolution and human evolution.

Genotype vs. Phenotype

When discussing evolution, it is important to distinguish between "genotype" and "phenotype," both of which play a key role in our understanding of evolution. Genotype refers to the particular DNA sequences or genes that an organism possesses, and it plays an important role in determining the phenotype or traits of that organism that are expressed. Thus, while the forces of evolution mainly act on phenotypes or observed physical traits, the traits that are selected for by evolution are passed on through the reproduction of their associated genotypes. However, not every gene will have its coded trait(s) expressed due to dominance displayed by certain traits, and this will be discussed further in the later chapters.

Darwin's Theory of Natural Selection and the "Survival of the fittest"

Darwin theorized that natural selection, or how a population or species changes over time to adapt to its environment, is one of the main driving forces behind evolution and the formation and differentiation of species.[60] In general, adaptation refers to any instances or forms of evolution that allow organisms to better survive in their environment and outcompete other organisms in that environment.[61] Some examples of traits which would increase the chances of survival include anatomical structures that allow for the consumption of otherwise non consumable food (such as specialized teeth that can break hard nuts) or provide protection from predators (such as a hard shell or spikes). Competition

58 Sloan, 2019
59 Sloan, 2019
60 National Geographic, 2019
61 National Geographic, 2019

refers to the fight or struggle over limited resources in an environment that usually leads to harm or death for all organisms involved. There are two major types of competition in ecology: interspecific competition (competition over resources between different species) and intraspecific competition (competition over resources and mates within a species). Evolution includes all forms of adaptation as well as other changes due to factors such as random mutations.

Every environment provides certain selective pressures, or factors that favour certain phenotypes over others. An example of a selective pressure which has affected and continues to affect the human population is the differing levels of sunlight across the world. Around the Earth's equator, where sunlight is the most intense, there is a selective pressure for people with darker skin (an advantage) compared to lighter skin (a disadvantage) as lighter skin is more prone to sunlight-related damage. Overall, it is unlikely that significant changes would occur at a steady rate in organisms if there weren't any environmental or selective pressures favouring the survival and inheritance of certain traits over others. Thus, the main force behind natural selection can be described as "survival of the fittest" as organisms possessing traits that allow them to better survive in their environment and circumstances at that time will be more likely to survive than other organisms. However, simply surviving without producing offspring is not enough to pass on one's unique traits to future generations. In other words, only the traits of organisms who both survive and successfully reproduce will likely be passed onto future generations and determine the traits of those future generations. Evolutionary fitness encompasses an organism's ability to both survive and successfully reproduce.[62] In other words, an organism's evolutionary fitness is determined by both natural selection and sexual selection, both of which will be discussed later in the chapter.

At any given time, there will be genetic variation within a species or population. This variation leads to the presence of many different traits and characteristics, with some being more suitable to the environment at the time. Organisms with traits that allow them to better survive or outcompete other organisms (according to the selective pressures in the

---

62 Than, 2018

environment) will be more likely to reproduce and pass on their traits to future generations.[63] Over time, unsuitable traits will die out from the species due to a lack of reproductive success while beneficial traits will become more and more prevalent within the species. Overall, natural selection states that organisms with higher evolutionary fitness will be more likely to survive and reproduce, thus making it likely that their unique traits will be expressed in future generations and may have a large presence depending on how beneficial they are.

Darwin's Theory of Sexual Selection and the Importance of Sexual Reproduction

Sexual selection refers to the various processes and factors behind a species' mating system, choices and rituals.[64] There are various factors that must be considered when choosing a mate, because they will be involved in the rearing of any potential offspring and their genes and traits will likely be passed on to the offspring in some form. Thus, it would be beneficial to mate with another member of the same species who possesses traits that would increase your offspring's fitness and would likely help you in raising your offspring.

Mate choice, or intersexual selection, refers to the selection of a mate based on factors that make them "attractive."[65] The criteria or how selective one is when it comes to choosing a mate is called mate bias. Mate bias is an evolutionary mechanism that usually increases the fitness of a species as only members with favourable traits will likely mate and pass on their traits. There are five major mechanisms that play a role in mate bias: phenotypic benefits, Fisherian or runaway selection, genetic compatibility, indicator traits and sensory bias.[66] These five mechanisms are criteria that an "ideal" or good mate would possess, as mates can provide both direct and indirect benefits to their partner. Direct benefits are advantages provided to the mate (such as material support and protection) while indirect benefits are advantages provided to offspring:

63 Than, 2018
64 Than, 2018
65 Macnow, 202
66 Macnow, 202

(such as better survivability and attractiveness).[67] Overall, genetics play a huge role in determining the phenotype of traits that are assessed during sexual selection, and therefore have a significant impact on whether an organism is able to mate and pass on its unique genetic information to future generations.

First, phenotypic benefits refer to observable traits that are attractive in a potential mate, such as a tendency for nurturing or protecting one's mate and offspring. Next, Fisherian or runaway selection refers to the sexual attractiveness of a trait that has either no effect on survival or has a negative effect, with a prominent example being the plumage of male peacocks. Female peacocks find the plumage attractive even though it comes as a disadvantage to the male's survival. Moving on, genetic compatibility refers to the attraction to other members of the same species with a different genetic makeup. Having a mate who has a significantly different genetic makeup would decrease the chances of producing offspring with homozygous alleles for genetic conditions (decreasing the chances of offspring having those genetic conditions), for example. If an organism is Next, indicator traits are traits that signify good overall health and wellbeing, and include genetic factors as well as indicators of disease or malnutrition, for example. The purpose of this mechanism is to ensure mating with healthy partners to increase the survivability of offspring. Finally, sensory bias refers to traits that match pre-existing preferences in the species. For example, male fiddler crabs often build uneven structures around their territory to attract mates by taking advantage of the natural tendency for fiddler crabs to be attracted to structures that appear uneven in the horizon.

Other important aspects of sexual selection are mating rituals and competition over mates.[68] In many animal species, males are either judged by females for their performance in mating rituals or courtship displays or must directly fight other males for mating rights. A mating ritual or courtship display is a unique behaviour or act in certain species which males must perform to attract a female.[69] Some examples include male

---

67 Macnow, 202
68 Birkhead, M. & Birkhead, T., 1990
69 Birkhead, M. & Birkhead, T., 1990

peacocks showing off their plumages to female peacocks and unique dances performed in some species of birds. Next, there is often fierce and violent competition for mates and the right to reproduce in many species. In many animal species, males compete in clashes of power and durability right to mate with females of their species. An example would be the clash of horns between male mountain goats during mating season, with the victors claiming the right to mate and pass on his traits while the losers fail to do so and become evolutionary "failures" in the process.[70] As will be discussed later in the chapter, genetics have an impact on mating rituals and customs as they play a role in determining the physical/anatomical structures (such as horns and increased muscle mass) and motivation (such as the hormones and pheromones that induce aggression) behind mating rituals and competition.

Finally, it is important to discuss the impact of sexual selection itself on evolution. While sexual reproduction does serve as a method of passing on genetic information to future generations, the information passed on is not an exact replica of the parents' genetic information. There are several steps in the process of forming gametes (meiosis) and sexual reproduction that allow for mixing and recombination of genetic material.[71] These mechanisms exist to provide further genetic diversity in a population or species and have the potential to create new sequences and traits that may be beneficial, especially if environmental circumstances change.

Behavioural Evolution and "Genetic Memory"

Many behaviours and skills, such as riding a bicycle, are learned through teachings and the examples set by others around us. However, there are many untaught behaviours that we still know how to do because they are coded within our genetic information. These behaviours are often referred to as "instincts", and they are performed unconsciously and cannot be changed by learning or external influences.[72] Thus far, we have discussed evolution mostly in terms of observable physical traits, but behaviours are also traits that also fall under the category of pheno-

---

70 Birkhead, M. & Birkhead, T., 1990
71 Wong, 2021 (Chapter 1)
72 Breed, 2010

types. Certain behaviours are coded within an organism's DNA and/or are affected by physical traits, meaning that behaviours can be heritable and subject to evolution just as physical and anatomical traits are.

Behaviours are closely related to and determined by physical traits such as anatomical structures, sensory systems and brain structures. Behavioural evolution usually occurs in 4 major areas: response to environmental stimuli, motivation, template and machinery to perform.[73] First, behaviours can be considered responses to sensory stimuli that are sensed and processed by an organism. Changes in sensory organs (that would change the detected stimuli) and the way that sensory stimuli are processed would alter behaviour. Next, motivation refers to the likelihood or probability of performing a behaviour. Motivation for a certain behaviour depends on certain brain structures, meaning that motivation would be subject to genetics and evolution as well. "Template" refers to the biological "blueprints" for a behaviour. The templates for most behaviours are usually determined by genetics. Finally, "machinery to perform" refers to the anatomical structures that allow a behaviour to be performed. For example, our hands allow us to manipulate and investigate objects in our environment.

Finally, the concept of "genetic memory" refers to the possibility that the memories or experiences of our ancestors could be passed on through genetics.[74] This concept has been somewhat popularized throughout pop culture, with some examples including the Assassin's Creed franchise, where the player plays as a person who relives the past experiences of his ancestors with the help of a specialized machine. Although more research needs to be done on whether ancestral memories are as readily accessible as suggested in pop culture, DNA provides the basis for many behaviours and could potentially contain some form of ancestral "memories" or remnants.

Inclusive Fitness
In 1964, William Donald Hamilton proposed two metrics of measur-

---

73 Sheehan et al., 2018
74 Gallagher, 2013

ing evolutionary success: personal/direct fitness and inclusive fitness.[75] Personal fitness refers to the number of offspring that an organism produces (regardless of who raises them) while inclusive fitness refers to the number of offspring "equivalents" that an organism rears, rescues or supports. As discussed previously, an organism's direct offspring will contain half of each parent's genetic information. The offspring produced by a sibling would have one quarter of one's genetic information, and would be considered half of an offspring equivalent, for example.[76] Overall, the offspring equivalent measures how related an offspring of a relative is, and the more offspring equivalents supported or raised, the better an organism's inclusive fitness is.[77]

The concepts of personal and inclusive fitness help to explain the instinct to nurture our children and protect and support relatives, whether it be our cousins, nieces, or brothers. This fundamental instinct directs us to protect relatives who share some of the same or similar genes to us. In a sense, protecting one's relatives and their offspring can help to ensure their survival and potential to reproduce, which would enable the passing on of genes similar to yourself onto future generations.

Artificial Selection

As discussed previously, the main driving forces of evolution include non-random forces (such as natural selection) and random occurrences. However, what would human influences on evolution be classified as? The results of human interference in evolution can be seen from modern agricultural crops that would most likely not exist otherwise to the emergence of dog breeds that would most likely not be able to survive on their own in the wild. From selective breeding to genetic modification, human attempts to divert evolution in a favourable direction (often when domesticating plants and animals) is referred to as "artificial selection."[78] It is a form of selection as certain traits are favoured and it is called "artificial" because the traits favoured by humans may not have been favoured under natural circumstances. There are two main

---

75 Szala & Shackelford, 2019
76 Szala & Shackelford, 2019
77 Szala & Shackelford, 2019
78 Gregory, 2009

types of artificial selection: methodical and unconscious.[79] Methodical selection occurs when humans deliberately attempt to modify a species according to predetermined standards while unconscious selection occurs when humans preserve desired or benefits traits that occur naturally while disposing of unwanted traits without the express purpose of altering the species.[80]

Even before biotechnologies such as genetic engineering were invented, humans have engaged in selective breeding throughout history, most significantly in terms of the selective breeding of agricultural crops and domesticated animals and livestock. The general process for selective breeding involves identifying plants or animals with desirable traits and having them mate with each other. Over time, this leads to the continuous reproduction of the desired phenotypes and behaviours. Essentially, human influences and desires play the role of selective pressures (which would usually be provided by the organism's environment) due to the control offered by domestication.

Recent advances in biotechnology have allowed for the faster evolution or development of desired traits with more control/precision compared to traditional selective breeding practices. In other words, the efficiency and effectiveness of methodical artificial selection has been increased. After all, humans have limited control over traditional selective breeding with random mutations and randomization of genetic information during the sexual reproduction process. However, modern biotechnology is not without controversy, as numerous concerns about its safety and potential legal and ethical issues have been raised. Regardless, selective breeding and biotechnology serve the same fundamental purpose of directing evolution in ways that benefit humans, and biotechnology opens up many exciting possibilities for the future of artificial selection and genetics.

79 Gregory, 2009
80 Gregory, 2009

3

The History of Genetics
Written By: Hanna Redda

Ancient Theories

While the finer details of genetics were not figured out until recently in scientific history, acknowledging that there are still aspects which scientists have not fully gained an understanding on,[81] there are signs that in ancient times our ancestors also shared this curiosity for genetics and inheritance. Scientists believe that in ancient times what sparked the desire to have a better understanding of genetics was the fact that people tend to take after their relatives. It was more than just the facial structure, in their external appearances people take after their genetically related relatives in many ways such as body shape, the pitch of voice, hair, skin colour and so many other traits. Historically, once the basics had been understood, scientists began breeding animals to test their understanding. The careful breeding techniques helped strengthen the livestock they had and thus improved their own stability as they were dependent on the success of the livestock.[82]

---

81 Winchester, 2020
82 Winchester, 2020

The now famous ancient Greek physician named Hippocrates had his own ideas surrounding genetics and inheritance.[83] Hippocrates believed that all valuable parts of the human body produced seeds and that those unseen seeds were what helped to create a child. According to Hippocrates' belief, during intercourse men would send these seeds and once met with a woman's seed the process of the development of the fetus would start. Even today, the term "seed" is used. Someone can refer to their offspring as their seed and most people understand the use of the word in that circumstance. It is interesting to note the influence of people from the past on today's people, and the culture and language in these modern times.[84]

Aristotle was one of the earlier proponents of the importance of blood in the human body and its relation to inheritance.[85] In his belief system for fully grown people, generative materials well-being was made possible due to blood being in the body. Aristotle believed that it was the actual blood of a person which was being passed down to the offspring. If you're wondering how one would physically give their blood to an offspring, since there is no exchange of blood process involved in the act of creating and carrying a child, Aristotle's explanation was that semen was a clean version of blood. He also thought that a menstrual cycle was the point at which a woman would supply her offspring with blood. Through the merging of blood, a fetus was developed and a baby was soon to be born. In essence, he considered blood to be the medium through which the future child would inherit traits from their parents. Modern science has disproved that blood is not at all a part of the baby-making process, however it can be noted that the term "blood" is often used to describe someone's kin. This can be attributed to Aristotle's continuing influence despite his passing many, many years ago.[86]

Classical Genetics

Classical genetics have laid the groundwork for the study of genetics, in fact it forms the foundation of genetics and is still the base of how genet-

---

83 Winchester, 2020
84 Winchester, 2020
85 Winchester, 2020
86 Winchester, 2020

ics are studied.[87] Classical genetics has a system of figuring out which genetic trait will be inherited by the offspring from the parents. Generally, if there is a dominant allele and a recessive allele, the dominant gene will be the trait that is inherited. There are other factors to consider, such as a certain chromosome through which the gene is passed down. Many of the principles in classical genetics have come from Greogor Mendel's study of inheritance with peas. In modern times the use for classical genetics is to uncover genes and locate the genes to then further study them at a structural level.[88]

As the subject of classical genetics was on the rise, experiments to better help the understanding of inheritance were being done. When conducting these experiments, the subjects being experimented on had to be chosen wisely.[89] Within some species, such as larger mammals, do not produce offspring often and when they do produce offspring there are not that many of them. These circumstances guided scientists to choose their subjects wisely, ideally species that have offsprings often and many offsprings at once. This led to scientists conducting genetic inheritance studies on species such as fruit flies.[90]

Genetic Disorders

At the root of medical genetics are diseases and inherited disorders. Kenelm Digby, was an English physician in the 17-century and the first person to discover an inherited abnormality.[91] The condition that was discovered is currently referred to as autosomal dominant inheritance disorder. The trait that Digby noticed was risk-free and did not pose a threat to the person's life or to the quality of their life. Digby found that a family had a "double thumb," a trait that had been found in five generations and had occurred in the women of the family.[92]

---

87 Winchester, 2020

88 Winchester, 2020

89 Winchester, 2020

90 Winchester, 2020

91 Harper, 2008

92 Harper, 2008

Albinism

The first autosomal recessive inherited genetic disorder to be discovered was albinism.[93] It was discovered by Lionel Wafer. Albinism is a skin condition associated with another condition such as vitiligo, where segments of skin do not have pigmentation. Albinism can be considered more disadvantageous to animals living in the wild compared to humans as affected animals may not be able to blend into their habitat or camouflage as well as others. This could lead to them having a shorter life expectancy (as it is easy for predators and prey to spot them, making it more difficult to escape predation and catch prey for food).[94]

In humans, there are two major types of albinism. There is Oculocutaneous albinism, which is where the albinism affects the pigmentation of a person in areas such as their hair, skin and eyes.[95] This type of albinism can be further subdivided into four subdivisions. People with this type of albinism have very light skin that sometimes may appear to be a pink-like colour and is due to the blood vessels showing from underneath the skin. They should take precautions when it comes to sun exposure. People with very light skin are more likely to burn their skin upon sun exposure, which puts them at a higher risk for developing skin cancer.[96]

Another one of the main types of albinism found in humans is called Ocular albinism.[97] This type affects the pigmentation of someone's eyes. In both Oculocutaneous albinism and Ocular albinism, the people affected will have very light eyes. The iris of the eyes will look to be a pink like colour and that is due to the fact that there is blood in the body that can be seen due to the lack of pigmentation. The pupils of the eyes will appear to have a red like colour for the same reason as the iris. Oculocutaneous albinism is caused by a genetic alteration that leads to there either being a decrease in melanin production of none at all.[98]

---

93 Harper, 2008
94 Harper, 2008
95 Harper, 2008
96 Harper, 2008
97 Harper, 2008
98 Harper, 2008

Colour Blindness

The first X related inherited genetic disorder is Colour Blindness.[99] Colour Blindness was discovered by John Dalton. Colour Blindness is described as the inability to differentiate between one or more of the following colors: blue, red, and green. People who have colour blindness do not have a strong colour sensing system. In contrast to people who have general blindness, people with blindness do not have a colour sensing system.[100]

Humans typically have three layers of cones in the retina.[101] One layer of the cone's purpose is to absorb the blue and violet wavelengths. The other layer of the cone's purpose is to absorb the green wave lengths. The last layer of the cone has the purpose of absorbing the red wavelengths. The cone that is supposed to observe the red wavelengths is supposed to be the most sensitive. There are a few different types of colour Blindness, including Dichromacy and Monochromacy. Dichromacy is when two of the three layers have the ability to operate. Dichromatic people are usually not able to tell the difference between the colours green and red.[102]

People who are unable to see the colour red have a blindness and a condition known as Protanopia.[103] Protanopia is when the cone layer that is supposed to absorb the wavelengths of the colour red are not working as they typically would, making it such that green and blue light are the only ones able to be absorbed. Individuals who are unable to see the colour green have a blindness to it and a condition known as Deuteranopia. That happens when the cone that should have been able to absorb the colour green, but the colours red and blue are still able to be absorbed and thus seen. When someone is unable to see the colour blue, but they can see red and green they may have a condition known as Tritanopia. Tritanopia happens when the cone layer that is traditionally supposed to absorb the blue wavelengths is not there. Monochromacy

---

99 Britannica Editors, 2017
100 Britannica Editors, 2017
101 Britannica Editors, 2017
102 Britannica Editors, 2017
103 Britannica Editors, 2017

is when one or zero of the cone layers are able to successfully operate. This results in a person's inability to see and differentiate any colour. Sometimes when someone has Monochromacy, it can also cause them to have a weakened ability to see overall.[104]

Red-green colour blindness that has been inherited has predominantly been found Caucasian men.[105] This type of colour blindness is present in roughly eight percent of Caucasian men. In Caucasian women, it is found at a rate of roughly half a percent. Colour blindness is an inherited disorder that is sex-linked and found on the X chromosome. Men are more likely to have colour blindness because sons inherit their X chromosome directly from their mothers. This means that when a mother is a carrier of a genetic disorder that is found in the X chromosome; it does not matter whether or not the father is a carrier or not because the father is only providing the Y chromosome. Basically, if a mother is a carrier of the gene the son will have it. However, if a daughter were to have a carrier mother and a father that was not a carrier of the gene she would not have due to her father providing the dominant (non-carrier) gene. Colour blindness is a disorder that is sex-linked and found on the X chromosome but it is also recessive meaning that when a dominant gene is present it will be the one that is passed on. In this case the dominant gene is the one that allows for people to see colour.[106]

Blue-yellow colour blindness is a dominant autosomal genetic disorder.[107] This particular type of colour blindness does not have a link to sex. Due to this gene being dominant, it does not help avoid it even if a person has a parent that is a non-carrier. All it takes to have the disorder is one parent being the carrier. Colour blindness that has been acquired and not present with the person from birth is predominantly found to be the blue-yellow type. The degree of acquired colour blindness can vary as some people with the condition are able to see colours to a limited extent while others cannot see colours at all. When the disorder is acquired it is usually caused by another disorder, an example would be

---

104 Britannica Editors, 2017
105 Britannica Editors, 2017
106 Britannica Editors, 2017
107 Britannica Editors, 2017

glaucoma or diabetes.[108]

Timeline of Milestones for Genetics

In 1866, Gregor Mendel made his findings on pea plants available[109] and his work later became extremely important to the study of genetics. In 1869, Johann Friedrich Miescher, a Swiss biochemist was the first person able to isolate what is now known as DNA. In the early 1900's Mendel's findings were reproduced by two researchers. This marked the beginning of modern genetics. In 1928, a British bacteriologist Frederick Griffith found out that bacteria have the ability to transfer genetic information and that genetic information was then able to be inherited. In 1931, Harriet B. Creighton and Barbara McClintock released a paper showing that some genes are connected with physically traded chromosomes. This study implied that chromosomes are the base of genetics. In 1944, Canadian bacteriologist Oswald Avery, a Canadian bacteriologist, Maclyn McCarty and American biologist Colin MacLeod discovered and reported that the genetic material found in the cell was DNA. In 1950, Erwin Chargaff, an Austrian biochemist, found that DNA's three components are equal in ratio. In 1951, Rosalind Franklin, Maurice Wilkins and Raymond Gosling guided studies with X-rays. This study allowed for them to find and capture the DNA fibers.[110]

In 1953, further advancements were made using the information that Erwin Chargaff found using X-rays. James Watson and Francis Crick were able to find the molecular structure of DNA[111] for which they received a Nobel Prize in 1962. In the 1970's American molecular biologists named Allen Maxam and Walter Gilbert, and a British biochemist named Fredrick Sanger came up with and advanced some of the very first methods that allowed for DNA sequencing. In 1982, Kary Mullis an American biochemist came up with PCR, which is also known as polymerase chain reaction. Polymerase chain reaction is a method that is easy to follow and makes it so that some patches of DNA can be copied billions of times in a short span of time. In 1993, Mullis was the recipient

---

108 Britannica Editors, 2017

109 Winchester, 2020

110 Winchester, 2020

111 Winchester, 2020

for the Nobel Prize for chemistry. From the 1990's to 2003, the Human Genome project began and successfully ended. By the end of this time period, scientists had the ability to have the sequencing of most all human genomes.[112]

---

112 Winchester, 2020

Gregor Mendel: The Father of Heredity & Mendelian Inheritance

Written By: Nataliya Raza

Introduction

Gregor Mendel assumed many roles throughout his life: monk, teacher, botanist, beekeeper and experimentalist. However, his most notable work is in the field of Genetics. His unprecedented discoveries expanded the knowledge of scientists and biologists alike and laid the stepping-stone for the development of an entirely new discipline. He was given the title the Father of Heredity, his work only recognized and cherished after his passing. Mendel was not a traditional 'scientist,' today he is remembered as a 'gentle man who loved flowers,' a priest and a self-effacing visionary.

The Biography of Gregor Mendel

Born to a German speaking family in modern-day Czech Republic in 1822, Mendel spent the greater part of his childhood on his family farm with his two sisters. His family strongly emphasized the importance of education despite their lack of financial resources. In his youth, he worked various roles as a gardener and a beekeeper.[113] Mendel was given his first window of opportunity as a child when his local priest

---

113 Scoville, 2019

recognized the unbridled potential in him at the ripe age of 11 and recommended his parents send him to school. Upon excelling in school, Mendel enrolled in a philosophy program at the Philosophical Institute of the University of Olmütz in Czech Republic in 1840. Despite his relative academic success in the fields of physics and philosophy, Mendel's parents struggled to support him financially and so he took to tutoring other students to ensure his expenses were paid for.[114]

Mendel's work began to take a toll on his mind and body where he was soon forced to take a break to recover from severe Depression.[115] He was sent home to recuperate but was also faced with challenges at home. Mendel's father desired for him to manage the family farm as the only son however, Mendel sought other ambitions. Upon graduating in 1843, he joined a Monastery as a means to eschew his father's request. As a new member of the Augustinian Abbey at the St. Thomas Monastery in Brno, he was conferred the name, Gregor.[116] The monastery served as a hub of culture, philosophy and science wherein Mendel first thrived off of.

However, with the passing of time, Mendel was expected to assume the traditional duties of a Priest such as visiting the ill and it distressed him greatly. Due to his inability to successfully manage this aforementioned duty, Mendel was sent to fulfill the short-term role of a teacher in Znaim. He had seemed to find both his talent and passion in teaching and was soon considering teaching as a full-time occupation. Mendel's passion fell short when he failed his teaching certification in 1850 and the Monastery decided he would benefit from two years of pursuing sciences at the University of Vienna. With the Monastery covering his tuition, Mendel worked under remarkable scientists and flourished, ultimately returning to teach at a secondary school. During his tenure at the school he began to experiment with the breeding of various plants in the Monastery's garden. In the year 1854, Mendel was allowed to conduct an experiment in the Monastery that would examine the distribution of traits in successive generations of hybrid plants. Mendel's cross

---

114 Olby, 2021
115 Olby, 2021
116 Olby, 2021

breeding experimentation would later prove to be one of the most profoundly insightful experiments in society and would serve as the basis of modern-day genetics.

Mendel's Experimentation

Mendel's journey to discover the science behind the patterns of inheritance first began in 1856. He initially experimented with the breeding of mice and then switched to honeybees with which he had little success, finally moving to peas as his preferred mode of variables.[117] These edible peas, known as Pisum sativum, served as an ideal model system, an organism that allows scientists to answer specific research questions and then make reliably broad generalizations based on their findings such as more complex organisms like humans. Peas were characteristically an ideal plant due to their 'numerous distinct varieties, the ease of culture and control of pollination, and the high proportion of successful seed germinations.'[118] The breeding of peas also required minimal attention and resulted in rapid growth. Lastly, due to having both female and male reproductive anatomy peas were unique in their ability to either cross-pollinate or self-pollinate.[119]

Mendel classified 7 uniquely distinct traits to observe with time, including plant height (tall or short), flower colour, seed colour (yellow or green) and shape (round or wrinkled). Peas were generally known to only exhibit one of the two dichotomous variables wherein a pea would only be short or tall, or yellow or green. These marked traits were known as character pairs. Mendel decided to breed one character pair across multiple generations to ensure the offspring always duplicated in the same manner producing the identical trait as the parent. The purpose of this pure-breeding was to ensure the dependability of his results and to yield reliable conclusions. He then began to cross breed one pure trait with another, such as a tall pea plant with a short pea plant. It was soon apparent to Mendel that one trait would successfully overtake and mask the other in the first generation of the hybrid (F1) offspring. If a tall and short pea plant was bred, the results would consistently yield a

---

117 Khan Academy, 2021
118 Olby, 2021
119 Scoville, 2019

tall pea plant. This observation spurred the concept of dominant and recessive traits wherein one was outwardly visible and the other one was hidden.[120] As Mendel continued to the second generation of plants (F2), he allowed the plants to self-fertilize and observed the recurrence of the initially hidden or recessive trait. However, the frequency of the hidden trait was far less than the dominant trait, a pattern that could be best described as a 3 to 1 ratio in which for every 3 tall pea plants one short pea plant occurred. Mendel tested and observed this distinct 3 to 1 ratio with other organisms and found the results to be similar. By 1865, Mendel had tested and analyzed more than 30,000 pea plants and had devised a pattern of inheritance. Additionally, Mendel noticed that the seven traits transmitted independently of the others, meaning a plant's height did not necessarily affect the colour of the flower or the seed.

The Law of Segregation

Mendel's systematic observation of cross-production prompted the formation of the widely known Law of Segregation. Prior to his work, the most widely prevalent theory surrounding genetics was the Blending Inheritance theory. Blending Inheritance purported children received permanent heritable information from both parents. However, there was no specific hypothesis, just a generalized theory which failed to account for the distinct differences observed between F1 plants wherein tall pea plants crossbred with short pea plants consistently produced tall pea plants. According to this theory, a tall pea plant and a short pea plant should produce a blended offspring, ideally a medium sized plant.[121] In both humans and plants, height is now believed to be an inheritable trait that is derived from the combination of unique specifically coded genes.

Mendel's observation with F1 pea plants revealed that parents transmit specific heritable material to offspring which is now classified as genes. Each gene carries a copy of a trait from each parent, resulting in two different versions known as alleles. If (T) represents a tall pea plant and (t) a short pea plant, the (T) tall allele may dominate the short (t) allele and result in the expression of an ultimately tall pea plant. Plants can either have a genotype that is homozygous or heterozygous and yet ex-

120 Khan Academy, 2021
121 KhanAcademy, 2021

press one distinct phenotype as alluded to in Chapter 2. However, only one of the two alleles in an organism is randomly assigned to a gamete.

Upon the pairing of a sperm and an egg, each gamete brings a unique allele that contributes to the genotype of the organism and is expressed phenotypically. The pairing of one specific trait such as height is categorized as monohybrid cross wherein only one trait is transmitted to the offspring.[122] Since traits could only be observed when they were finally expressed, it was difficult to determine the genotypes of a specific organism. Mendel successfully overcame this obstacle by developing the test cross. The test cross is a technique used till his day and requires the crossing a dominant phenotype (could be either homozygous or heterozygous dominant) to be bred along with a homozygous recessive of the same trait. This procedure could determine whether the phenotypically tall pea plant was either (TT) or (Tt) by breeding it with a short pea plant (tt).

According to the possible results and our understanding of genetics through punnett squares to be discussed in Chapter 5, if all of the offspring were tall it would indicate the unknown genotype must have been homozygous dominant (TT). Although if the offspring yielded a mixture of both short and tall pea plants, the pea plants would comprise of both (Tt) and (tt), that would indicate a heterozygous dominant genotype (Tt) for the unknown parent's genotype. Test crosses helped Mendel solidify and verify his hypothesis and build a reliable set of data.

Law of Independent Assortment

As previously alluded to in Mendel's experimentation, Mendel had discovered that each trait has unique alleles which sort independently of each other. This means that an allele of one distinct trait is randomly assigned to a gamete independent of another allele assigned for a different trait.[123] If we examine genetics in humans, the height of a human is assigned independently of the individual's eye colour. These two traits have no direct correlation and do not influence each other.

---

122 Scoville, 2019
123 KhanAcademy, 2021

In contrast, Mendel's peas also revealed that a pea plant's alleles for a designated trait are not influenced by another trait. For instance, the height of a pea plant is not governed by the shape of its seed. As such, a tall and round seeded pea plant could hypothetically be categorized as (TTRR) while a short and wrinkled plant would be classified with the (ttrr) genotype. If one was to crossbreed these two, based on the homozygous nature of each the results should be (Tt) for height and (Rr) for seed shape, which would eventually yield (TtRr). Since the allele for tall and rounded seed shape is dominant, the recessive homozygous traits in one parent would get masked by the dominant parent resulting in a tall and round seeded offspring. Now this is what is known as a dicross hybrid that transmits not one but two traits across generations.

Based on relatively new information of chromosome structure and meiosis, Mendel's theory of Independent Assortment has been proven to be true. Genes occur in the form of alleles on various different chromosomes. The allele for plant height may be located on chromosome 2 whereas the allele for seed shape may be on chromosome 7. When these genes are assorted, they often assort independently. Moreover, it is possible for genes on the same chromosome to also assort independently of each other if they are a certain distance apart from each other. While Mendel was correct in his theory, there are certain genes that tend to assort together based on their proximity with each other. This often occurs in the case of linked genes which are alleles closely located beside each other and often assort as one cohesive unit.[124]

Mendel's Impact

Mendel ultimately crossbred and observed fifth, sixth generations and so on of pea plants, but his model of inheritance consisted mainly of the parent generation and F1 and F2. He continued his work for a number of years and finally presented it in 1865, however he gained little to no recognition during that time. His work was published under the title of 'Experiments in Plant Hybridization' the following year, yet it wasn't perceived as innovative enough. Many scientists now believe that it was too unconventional for its time and even eluded other notable intellectu-

---

124 KhanAcademy, 2021

als such as Charles Darwin.[125]

Mendel also believed that to be true and is remembered for saying that 'My time will come,' hoping people would someday recognize the potential in his work.[126] Mendel resumed his research activities and later gained more experience in the field of meteorological and apicultural work. He did not attempt to claim any type of publicity or awareness of his work in his latter years. His final years were spent travelling once to England to witness the Industrial exhibition with the majority of his time prior to his death was spent nurturing his health. He eventually passed away from Bright's disease and was mourned by his fellow monks and siblings.

Rediscovery by other Scientists

Mendel did not gain much acknowledgement until years later when several European botanists and geneticists independently discovered similar results in their experiments with plant hybridization. While few were mildly familiar with Mendel's work, they did not gain a comprehensive understanding until compiling their research together. It was the biologist, William Bateson who began to propagate Mendel's work across an increasing number of scientists and intellectuals. Their ideologies were believed to be contrary to Darwinism which claimed to be based on the selection of 'small blending variation' whereas Mendel's work focused on 'non blending variation.'[127] This aroused a great deal of enthusiasm and pride from Mendel's fans and thus were given the name 'Mendelians.'

It wasn't until 30 years later in which Mendel's Inheritance Model was deemed critical to the theory of Evolution. Although the concept and term, 'genes' was not determined until 1909 by botanist, Willhelm Johansen, Mendel's model remained the basis of the science of genetics for centuries to come.[128] His ingenious experimentation and comprehensive analysis was pivotal in the development of the field of genetics and the rudimentary understanding of life itself. It is due to this, he is

---

125 KhanAcademy, 2021
126 Nature Publishing Group, 2021
127 Olby, 202
128 Olby, 202

proclaimed to be the Father of Genetics.

Conclusion

Mendel's work laid the foundational framework not only for the field of botany or genetics but provided humankind with a glimpse of the very origins of life on a microscopic level, all the while demonstrating the profound implications it holds in our day-to-day lives. His understanding of heredity in centuries past has given us the opportunity today to clone sheep, decode the human genome and provide complex gene therapy with the potential of saving millions of people across the globe. Mendel might have been undervalued and unappreciated in his time, however today his work is a cornerstone in biology textbooks and he, a founding father in the science hall of fame. Mendel's Model of Inheritance lives on in newly discovered therapies, scientific breakthroughs, across plant farms and within our very own bodies.

5

What is the Genetics of Heredity?

Written By: Hareem Bilal

Genetics: The Study of You and Me

Most human diseases and traits have an associated genetic component or factor regardless of whether they are influenced by other factors such as a lack of exercise. The human body's response to environmental factors such as toxins can be modified by genetic components. To improve the application of genetic information, technologies, and disease diagnosis, it is necessary to have thorough understanding about the underlying concepts of human genetics, and roles of environment, genes and behaviour.[129]

The ability to reproduce is the most fundamental property in all living organisms. Parents pass down genetic information specifying function and structure to their offsprings. Similarly, cells are created from pre-existing ones, as genetic material in cells is passed from parent to progeny through replication. The central question of all biology is how replication and transmission of genetic material from organism to or-

---

129 Genetic Alliance, & The New York-Mid-Atlantic Consortium for Genetic and Newborn Screening Services, 2009

ganism and cell to cell takes place. The foundation of our current understanding of molecular biology is due to the in depth research on mechanisms of genetic transmission and the identification of genetic material as DNA.[130]

Genetics, the study of heredity, has advanced exponentially in the field of science, and is recognized as the "forefront of biological pioneering".[131] Genetics is remarkably the most qualitative of all biological categories. It is close to the physical sciences due to the application of mathematical principles and has "emerged the hierarchy of the established sciences."[132] The study of inheritance in organisms that are lower in the hierarchy was used to produce the current knowledge we know on genetics.[133]

Darwin's Incomplete Proposal

Charles Darwin formulated the theory of evolution by natural selection in the mid 19th century. Evolution is defined as the change in heritable traits of an organism over time, which contributes to diversity in the populations.[134] Darwin's theory of natural selection failed to logically provide an adequate explanation on the inheritance of traits. He could not fully explain the characteristics inherited during the life of an organism. Even though Darwin shared a probabilistic and mathematical outlook with Gregor Mendel, he could not interpret his own data showing Mendelian ratios, as he lacked a model that presented the mechanisms of inheritance.[135]

In order to generate variation, Charles Darwin believed it was necessary for environmental changes to act upon either reproductive organs or the entire body. Although, heredity for him was not a transmissional process, but a developmental one. In other terms, variation took place when the developmental process of change was affected by the envi-

---

130 Cooper, 2000
131 Crummy, 1942
132 Crummy, 1942
133 Crummy, 1942
134 Ashraf & Sarfraz, 2016
135 Charlesworth & Charlesworth, 2009

ronment. Darwin understood the importance of environment and the necessity of variation and adaptation. He proposed natural selection as the external mechanism for adaptation, and proposed environmental changes as an external mechnison for variation. He stated that hereditary was the result of the parent being induced with change: "The father being climatized, climatizes the child."[136] He also noted that not every change induced on the parent was heritable: "Any change suddenly acquired is with difficulty permanently transmitted."[137]

In 1844, Darwin expressed his views on the origination of variation in his essay, where some variation was attributed to "the laws of embryonic growth and of reproduction," but some variation was also caused by "the indirect effects of domestication on the action of the reproductive system" or an environmental action coined as germinally-mediated. "Considerable change from the natural conditions of life" causing reproductive organs to fail would result in germinally-mediated variation.[138] He never gave up the belief that disturbance in reproductive systems by external conditions caused variations. He believed in two kinds of externally-induced variations: somatically-mediated and germinally-mediated, or "direct" and "indirect". The "indirect" effect of variation was caused by environmental effects on reproductive organs of parents, which caused variation in the offspring, and was displayed in the offspring only, thus having an "indirect" effect. The "direct" effect of variation was caused by the environment acting on the organisms' body, causing and displaying variation in the same generation. The term "mediated" refers to the reproductive organs or body linking the environmental effect to changes in reproductive organs.[139]

Winther (2000) states the two internal mechanisms for variation proposed by Darwin: "(1) crossing between organisms of the same variety and (2) crossing between organisms of different varieties or species." Darwin proposed that, "[a] certain degree of variation (Muller's twins) seems inevitable effect of process of reproduction." Winther (2000) ex-

---

136 Winther, 2000
137 Winther, 2000
138 Winther, 2000
139 Winther, 2000

plains how Darwin vaguely discussed the first mechanism, and focused more on the second one. The crossing of different varieties or species was discussed on Chapter 8 "Hybridism" in his book On the Origin of Species (1859):

"The slight degree of variability in hybrids from the first cross or in the first generation, in contrast with their extreme variability in the succeeding generations, is a curious fact and deserves attention. For it bears on and corroborates the view which I have taken on the cause of ordinary variability; namely, that it is due to the reproductive system being eminently sensitive to any change in the conditions of life, being thus often rendered either impotent or at least incapable of its proper function of producing offspring identical with the parent-form. Now hybrids in the first generation are descended from species (excluding those long cultivated) which have not had their reproductive systems in any way affected, and they are not variable; but hybrids themselves have their reproductive systems seriously affected, and their descendants are highly variable".[140]

According to Darwin, changes in conditions of life or crossbreeding of different varieties or species affect reproductive organs. Even though he did not use them often, Darwin only allowed these explanations involving altered reproductive systems throughout his career. He noted in The Variation of Animal and Plants: "Although we have not at present sufficient evidence that the crossing of species, which have never been cultivated, leads to the appearance of new characters, this apparently does occur with species which have been already rendered in some degree variable through cultivation."[141] He believed variability in hybrids was due to the exposure of reproductive organs to changes in life conditions. Darwin made hybridization, his only internal mechanism of variation, dependent on the changes in environment.[142]

Mendel's contribution: Mathematical aid

Gregor Mendel was the first to apply calculus of ratios to biological sit-

---

140 Winther, 2000
141 Winther, 2000
142 Winther, 2000

uations, as he formulated a hypothesis and tested it on larger sets based on comparison of expected ratios and observed numbers. His mathematical approach can be attributed to his training in the physical sciences such as meteorology along with botany and plant breeding. One of his prominent innovations was that he viewed trait inheritance as random events and analyzed results based on expectations. He observed manifested traits to be "dominant" and hidden traits as "recessive"; terminology which is still used in modern genetics. Along with peas, he tested several other plant species such as Geum, Hieracium and Cirsium, which reflected a key question at that time, namely, how traits were transmitted after hybridization of species, in order to gain a deeper understanding on the origin of species. Instead of investigating the origin of species, Mendel looked for laws governing trait inheritance that did not change over time. He also rejected the theory of blending characters and species essence. Unlike Charles Darwin, who believed species varied over time, Mendel believed the characteristics of species remained constant. Chapter 4 describes in further detail about Medel's history and major contributions such as the pea experiment.[143]

Punnett squares and Mendelism

Born in 1875, Reginald Crundall Punnett finished his classical education and decided to work at Cambridge University with William Bateson in 1903. At Cambridge, he became the first professor of Genetics (1912), with expertise in genetics and poultry. He strongly supported Mendelism, a position stating that all observable characteristics are caused by genes that are directly transmitted through germ cells and are not acquired through experience. He wrote numerous papers on genetics, and worked together with Bateson and Miss E.L. Saunders to provide the earliest evidence for Mendelian inheritance. For all higher organisms, he saw clear implications of Mendelism by the quote in his book Mendelism:

" People generally look upon the human species as having two kinds of individuals, males and females, and it is for them that the sociologists and legislators frame their schemes. This, however, is but an imperfect view to take of ourselves. In reality we are of four kinds, male zygotes

---

143 Smýkal et al., 2016

and female zygotes, large gametes and small gametes, and heredity is the link that binds us all together.".[144]

Punnett devised a 3rd edition of Mendelism in order to think about the independent assortment of characters in gametes, and to compact all possible combinations of gametes which he called a "chessboard". This "chessboard" is known as the Punnett square (1950-1960); the name was later changed to honor and recognize Punnett's contributions.[145]

Heredity patterns can be shown through Punnett squares based on one, two or more gene pairs. A Punnett square represents the offspring resulting from different combinations of gametes fused in a cross.[146] It is a simple square shaped matrix, where genes or alleles of one parent are written on the row headings, while alleles of the other parent on the column headings. The focus of the discussion is on 2X2 matrices, but matrices can range in sizes depending on the amount of alleles studied at the same time. Possible outcomes of crossbreeding two parents are illustrated on the four cells in the matrix.

The offspring have either dominant heterozygous or homozygous genes, or recessive homozygous genes. If t represents green seeds and T represents yellow seeds, then the offspring (Tt) will have yellow seed due to yellow being the dominant trait. An example of second generation breeding (Tt x Tt) would result in four possible outcomes: "(1) a dominant homozygous offspring (TT), (2) a recessive homozygous offspring (tt), (3) a heterozygous offspring (Tt), or (4) another possibility of a heterozygous offspring (Tt)."[147] Three out of four times, the offspring will have the dominant characteristics since T dominates t.[148]

Punnett squares are used to systematically and economically visualize gamete combinations to create genotypes based on Mendel's theory. Punnett squares evolved and became a standard way of representing ge-

---

144 Davis, 1993
145 Davis, 1993
146 Griffiths et al., 2000
147 Brahier, 1999
148 Brahier, 1999

netic based problems. They became a central and classic tool for Mendelian genetics, and were adaptable to theoretical change.[149]

Genetic Variation in the Human Genome

Interactions between environmental and genetic factors play a vital role in disease production and its development throughout life. The genomic differences seen in the species or population is termed as genetic variation. Genetic variation is known as the parameter controlling an individual's phenotype due to the huge diversity in the human genome. This variation is present in various forms within the human genome, and throughout the genome, it occurs at different frequencies.[150]

Genetic variation is caused by recombination, mutation and immigration of genes. Any process in partly diploid or diploid cells that generate chromosomal combinations or new genes not found in its progenitor or in that cell is scientifically known as recombination. Unless alleles are segregating at different loci, recombination itself does not produce variation. For the same allele, if the entire species is homozygous, then immigration is unable to provide variation. As a result, all variation ultimately comes from mutation. Even though mutation is a source of variation, it does not drive evolution. Low rates of spontaneous mutations result in low rates of change in gene frequency from the process of mutation. The rate of mutation is defined as a probability of an allele changing to another from one generation to another. The increase in the rate of mutation is extremely slow and will get slower for every generation, as the older allele copies have yet to mutate.[151]

Recombination as a means of variation can be faster when compared to mutation. As an example, when two "normal" survival chromosomes from a Drosophila natural population are permitted to recombine for a single generation, they yield an array of chromosomes with 25 to 75 percent as much genetic variance in survival as the entire natural population from which the parent chromosomes were sampled. This outcome is the result of the exponentially large number of various recombinant

149 Winsatt, 2012
150 Talseth & Scott, 2011
151 Griffiths et al., 2000

chromosomes that can be produced, even if single crossovers were the ones taken into account. Variation through immigration which includes the movement of gene frequencies as one population travels to another. The resulting population will have intermediate allele frequency, somewhere between the original and the donor frequency.[152]

Population Genetics: Genetics and Evolution Hand-in-Hand

Understanding how the evolutionary process is driven by the mechanisms of natural selections and other factors, such as mutation and random drift, is the central goal of evolutionary biology. Questions are addressed qualitatively by population geneticists, as they build mathematical models, develop the statistical methods for inferring parameters of ancestral processes, and test formulated hypotheses which are based on the analysis of real data.[153]

The goal of population genetics is to understand the genetic composition of a population and the forces that change and determine that composition.The existence of various alleles at different gene loci in any species give rise to a great deal of genetic variation between and within populations. In population genetics, the frequency at which the alleles are found at any gene locus of interest is a fundamental measurement. Recurrent mutation, migration, selection or random sampling effects can change the frequency of a given allele. For a given locus, a randomly interbreeding population would show constant genotypic frequencies in an idealized population with no forces of change acting upon it.[154]

Population genetics uses the rules of inheritance to make predictions on how under the forces of evolution the genetic composition of a population will change, and compares relevant data to the predictions. Due to the increase in knowledge on the function and organization of genomes, population geneticists have to now confront the increase in the range of problems. Basic insight regarding the mechanisms of evolution is provided by population genetics shows its fundamental importance.[155]

---

152 Griffiths et al., 2000
153 Chen, 2015
154 Griffiths et al., 2000
155 Charlesworth, 2015

Two related problems gave rise to the indigestion of heredity by Mendel: how to understand the nature and origin of species and how to breed improved crops. The common factor in these problems and what differentes these problems from the transmission and gene action problems is that their concern does not lie with individuals but with populations. Studies concerning protein synthesis, development, gene replication and chromosome movement focus primarily on the process that goes on within the cells of an organism. However, it is important to note that change in the properties of a collective population or sets of populations takes place when a species goes through a transformation either by deliberate human intervention or through normal course of evolution.[156]

---

156 Griffiths et al., 1999

What is Population Genetics?

Written By: Vivek Kannan

What is Population Genetics?

Population genetics is a field within genetics that investigates genetic variance in populations. More specifically, allele frequencies are studied and mathematically analyzed to anticipate whether a single or group of alleles will appear in a population.[157] For context purposes, alleles are variations of a particular gene. For instance, if a specific gene within a flower codes for petal colours, the allele would be the variants of this gene, which would be the different colours that the petal can have (i.e., red versus pink petals).[158] Allele frequency is formally defined as how often an allele can be found in a population. For example, if there is an allele denoted as 'X,' one can find the frequency of allele X by dividing the number of allele X copies in a population by the total gene copies within the population.[159,160] The field of population genetics is particularly important to consider and study further into, because it provides a

157 "Population genetics," 2017

158 "Allele," n.d.

159 "Allele frequency," n.d.

160 Buckley, 2021

more clearer explanation as compared to previously proposed theories that stated how traits in species change over time (i.e., evolution).[161,162]

History of Population Genetics

Before diving into the specifics of population genetics, it is important to understand how it all started. The study of population genetics was established in the 1920-1930s by two British geneticists named R.A. Fisher, J.B.S. Haldane, and an American geneticist named Sewall Wright.[163,164,165,166] A question arises: what really led them to this profound subfield of genetics? It was, in fact, the theories of evolution proposed by Gregor Mendel and Charles Darwin.

Mendel theorized that a factor (i.e., allele in modern times) from each parent is passed down and formed as a pair in the offspring. In addition, he stated that there are alleles that are always expressed (i.e., dominant) and some that are only expressed when not paired with a dominant allele (i.e., recessive). On the other hand, Darwin heavily believed in natural selection where species adapt to their respective environments by favouring traits that give them the best chance of survival. An example of his supporters was biometricians, who utilized statistical analysis to investigate heredity. At this point in time, Mendel's theory was rediscovered in 1900.[167]

As with many theories, there were conflicting aspects with what Mendel and Darwin proposed. For instance, Mendel's theory supports the idea that discontinuous variation occurs during the evolution of species. This essentially means that species greatly change from parent to offspring, and the expression of traits is dependent on variations of a particular gene as opposed to a combination of genes. In contrast, Darwin's natural selection relied on the basis of gradualism, in which species change

---

161 Clark, 2003

162 "What is evolution?", 2017

163 Okasha, 2012

164 "Sir Ronald Aylmer Fisher," n.d.

165 "J.B.S. Haldane," n.d.

166 "Sewall Wright," n.d.

167 Okasha, 201

over time in incremental steps from parent to offspring. This theory follows a concept known as continuous variation, where traits are expressed as a result of a combination of genes.[168] In addition to this, Darwin was not able to explain a possible mechanism of inheritance for his natural selection theory and in fact, he explicitly stated this in his book: On the Origin of Species by Means of Natural Selection.[169] Thus, given the issues and conflicts that existed in both theories suggested by Mendel and Darwin, these theories were widely viewed as opposites of one another. As such, population genetics emerged, while scientists like Fisher, Haldane and Wright used mathematical models to merge the principles from Darwin and Mendel's theories.[170]

Firstly, Fisher contributed to the merging of theories by showing that a normal distribution curve of traits is possible in a population if a trait like the span of a hand (continuous variation) is altered by numerous genes, given that each gene has an effect on the particular trait. Under Darwin's theory, continuous variations like height and weight are taken into account and as such, it is expected to obtain a normal distribution for traits that are continuous. Typically Mendel's theory revolves around discontinuous variations, however, Fisher was able to obtain normal distribution curves that correlated with the trait distributions that followed Darwin's theory.[171,172] This result was significant as it helped combine both theories. Furthermore, Wright contributed by proposing that allele frequencies are subjected to change due to random chance, in which this phenomenon is more prevalent in smaller populations. This is also known as genetic drift.[173,174]

Interestingly enough, he stated that genetic drift can potentially lead to favourable traits in which he propounded that these ideal traits can only be present when large populations are divided into smaller, separated

---

168 Millstein & Skipper, 2007
169 Okasha, 2012
170 Okasha, 2012
171 Okasha, 2012
172 Millstein & Skipper, 2007
173 Millstein & Skipper, 2007
174 "Population Genetics," n.d.

populations, given that gene interactions between species of different populations is controlled (i.e., gene flow).[175,176] Eventually, Wright compressed these ideas and formulated his Shifting Balance Theory which states: "Evolution proceeds via a shifting balance process through three phases: Phase I, Random genetic drift causes subpopulations semi-isolated within the global population to lose fitness; Phase II, Mass selection on complex genetic interaction systems raises the fitness of those subpopulations; Phase III, Interdemic selection then raises the fitness of the large or global population".[177] Finally, Haldane supported the uprising of population genetics through his mathematical models, with an emphasis on natural selection. His mathematics prompted further research into migration and mutations, particularly how they influence natural selection. In addition, Haldane's work provided insights into gene frequencies and their rates of changes.[178]

The scientific work done by Fisher, Wright and Haldane helped resolve debates between Darwin and Mendel's theories by merging aspects of each theory together. This was accomplished using a quantitative approach via mathematical models and thus, translated into the beginning of population genetics.[179]

Factors That Affect Population Genetics

In order to truly understand population genetics, one must be aware of the different ways that can affect the allele frequencies in populations. There are five factors that affect population genetics and these include natural selection, sexual selection, mutations, genetic drift and gene flow.[180] Firstly, since natural selection is the process by which the traits that promote the greatest survival rate are passed on, the frequencies of the favourable alleles thereby increase while being passed on to future generations. A simple example of this can be seen in a predator and prey scenario. If there is a hawk that can only see white squirrels as opposed

---

175 Millstein & Skipper, 2007

176 "Gene flow," n.d.

177 Millstein & Skipper, 2007

178 Millstein & Skipper, 2007

179 Okasha, 2012

180 "Population genetics:," n.d.

to black squirrels, there will be a higher chance of survival for black squirrels. This means that the allele corresponding to the black colour trait will be passed on more frequently as opposed to the other colour as a result of more black squirrels reproducing.[181]

The next factor is sexual selection. Essentially, species will mate with those that are more attractive and display particular traits of interest (i.e., nonrandom mating). As such, more reproduction will occur and result in more offspring possessing the desired alleles.[182] Another factor that influences population genetics is mutations. These occur when there are errors on the DNA which can consequently be fatal, lead to deformations or be harmless to the individual. A few examples of mutations include insertions, deletions and nucleotide base substitutions.[183] Moreover, there are cases when mutations form new alleles that are favourable but harmless for a given species. For instance, a mutation might provide an advantage for food hunting or mating purposes. Therefore, when this occurs, these traits will be passed onto the offsprings and allele frequency will increase within the populations.[184] The other two factors are genetic drift and gene flow, which were explained above. An example of genetic drift is a randomly occurring event like a hurricane whereas gene flow can be demonstrated by the immigration of species where different genes can interact with one another.[185]

The Hardy-Weinberg Principle

When considering how allele frequencies change in populations while taking into account the different factors that affect it, one must understand that a very important principle known as the Hardy-Weinberg Principle was used to verify population genetics. This concept was first introduced in 1908 by an English mathematician named G.H. Hardy and a German physician known as W. Weinberg.[186,187] This key princi-

---

181 "Natural selection," n.d.
182 "Population genetics:," n.d.
183 "DNA Mutation and Repair," n.d.
184 "Population genetics:," n.d.
185 "Population genetics:," n.d.
186 Okasha, 2012
187 Millstein & Skipper, 2007

ple explains that allele frequencies will be constant and will not change between generations as long as the previously mentioned factors (i.e., natural selection, sexual selection, mutations, genetic drift and gene flow) are not present. This holds true under assumptions like having a large population and random mating.[188]

In fact, a population that resides in the state proposed by the Hardy-Weinberg Principle is known as the Hardy-Weinberg equilibrium. Furthermore, Hardy and Weinberg proposed an associated equation with this principle that follows: p2 + 2pq + q2.[189] This mathematical equation coupled with the general principle provides an insight into the relationship between the frequencies of the genotype and alleles. It also explains that natural selection is key for evolution, and considering only inheritance and genetic variation does not completely describe evolution as accurately. In addition, this principle provides a method for assessing the level of disturbance felt by the individual population genetic factors.[190]

Applications of Population Genetics

Currently, there are many applications that population genetics can be used in. For example, the Hardy-Weinberg Principle equation (i.e., p2 + 2pq + q2) is used when trying to determine whether an individual will be at risk of carrying a harmful allele. Furthermore, parts of the Hardy-Weinberg equation are used for statistics relating to DNA fingerprinting. More specifically, an individual is first tested if they are heterozygous or homozygous using DNA fingerprint patterns. Afterwards, the Hardy-Weinberg equation is manipulated to determine the genotype frequencies.

During the analysis step of a criminal investigation, one can figure out whether an individual is likely to be guilty or not. This can be accomplished by analyzing the groups of alleles. If the specific alleles match the ones in the crime scene as well as the individual's DNA, but knowing that the alleles are typically not found in the individual's population, it

---

188 Millstein & Skipper, 2007
189 Millstein & Skipper, 2007
190 "SparkNotes," n.d.

can be deduced that they are most likely guilty of the crime they committed.

Another application that population genetics is useful for is molecular clocks. The purpose of this is to find the approximate time when species diverged from their common ancestor. This is performed by cross-referencing the homologous proteins among the species of interest. Investigating these proteins is important because amino acid sequences can be further analyzed to identify differences (i.e., mutations) and subsequently determine the molecular clocks of the species.[191]

Closing Remarks

Population genetics is a unique field within genetics that deals with the changes in allele frequencies within a population. It is very important because it is based on the core principles outlined in the theories proposed by Charles Darwin and Gregor Mendel, and as such, population genetics is viewed as the clearest and in-depth explanation of heredity and evolution. In addition, population genetics can be better understood by being cognizant of the factors that affect this subfield include natural selection, sexual selection, mutations, genetic drift and gene flow. It is also important to note that the Hardy-Weinberg Principle is a fundamental concept that plays a crucial role in understanding how population genetics works. This field has greatly impacted the scientific community over the years. Given that a vast amount of technologies are being developed in this era, one can look forward to the possibility of further deepening the current understanding of genetics.

---

191 (Athira, n.d.)

7

What are Genetic Mutations?

Written By: Natalie Jean-Marie

Introduction

Mutations are defined as any heritable change in DNA sequence.[192] In this context, heritability refers to the ability to be passed on from parent to daughter cell, not necessarily from the parent organism to offspring. Mutations may happen during development or at any point in an organism's life. They are the basis for all genetic variation, as without mutations, natural selection could not occur. Changes caused by mutations are often neutral with regards to fitness, meaning they occur in non-coding DNA or contribute some small change, such as eye colour. Mutations may also have a negative impact on an individual's health or survival. This can range from minute disadvantages to fatal illnesses. On rare occasions, mutations are also beneficial to an individual. Depending on the type of cell affected by a mutation, the change may or may not be passed down through generations. To fully understand how mutations function, this chapter will describe the different types of mutations, how they occur, the mechanisms that prevent them, and how they are inherited. Finally, as mutations are often the basis of non-hu-

---

192 Morris, 2019

man creatures and mutants in pop culture, common misconceptions of mutations will be further explored.

Background

Before understanding mutations and how they impact an organism, there are a few concepts that need to be explored. The central dogma of molecular biology outlines the steps that take us from DNA to protein, which is the foundation for life. This has been outlined in Chapter 1, but now we must take a closer look, specifically at translation from RNA to protein. A sequence of three nucleotides is known as a codon. Each codon, then, is translated to a single amino acid, unless it is a stop codon, which signals to the translation machinery that the sequence is over.[193] Amino acids are the smallest unit that make up a protein. While there are many possible combinations of three nucleotides, there are only twenty resulting amino acids. This is due to redundancy in the genetic code. Most amino acids can be translated from roughly 3 different codons.

Types of Mutations

With such a complex system of DNA replication and translation, there are many opportunities for mutations to occur. As a result, there are several kinds of mutations, each with different causes, different processes, and different outcomes. This section will explain point mutations, chromosomal mutations, and copy number variations, as well as all of their subtypes.

A point mutation is a change in a single nucleotide or base pair.[194] Often, this is a substitution during DNA replication, which may or may not have any effect. For example, if the error involves replacing the last cytosine in the codon CCC with a guanine, the sequence would still code for the proline amino acid. Overall, there would be no change in the amino acid sequence and thus no higher-level changes. These types of mutations are referred to as silent or synonymous mutations.[195]

---

193 Morris, 2019
194 Clancy, 2008
195 Clancy, 2008

Alternatively, missense or nonsynonymous mutations occur when the nucleotide replacement does change the amino acid.[196] Missense mutations, even though they seem like a small change in our vast genome, have the ability to cause severe problems. Sickle-cell anemia, a disease that results in misshapen red blood cells, is caused by a substitution on the gene that codes for the beta chain of the hemoglobin protein. This substitution, which causes the amino acid glutamine to be replaced with valine, causes affected red blood cells to carry oxygen much less efficiently than healthy cells, and is often fatal for affected individuals.[197] Finally, substitution may result in a nonsense mutation, which creates a premature stop codon.[198] These are often the most harmful mutations, as an early stop codon may prevent vital proteins from being produced.

Other types of point mutations are insertion and deletion. Any number of nucleotides may be added or deleted, but these mutations tend to have the largest impact when the number is not in sets of threes. Nucleotides are read in sets of three to form amino acid, so an insertion of one nucleotide, for instance, causes a frameshift, which entirely changes the amino acid sequence.[199] This can prevent the proteins produced from performing their function, as is the case with cystic fibrosis. It is important to note that while most given examples have been diseases, point mutations are not always harmful. In fact, single nucleotide polymorphisms (SNPs) are single base pair alterations that occur in more than 1% of the population, majorly contributing to genetic variation.[200] They may occur in non-coding sequences of DNA and have no effect, or cause minor phenotypic differences, such as curly hair.

The largest scale type of mutation is chromosomal mutation, which refers to a change in large sections along a given chromosome. There are four main possibilities, which each affect the individual to varying degrees. An inversion refers to when a segment of DNA is detached, re-

---

196 Clancy, 2008
197 Clancy, 2008
198 Morris, 2019
199 Clancy, 2008
200 Clancy, 2008

versed, and reinserted into the chromosome.[201] This can cause diseases like Opitz-Kaveggia syndrome. Caused by several mutations on the X chromosome, Opitz-Kaveggia syndrome causes intellectual disability, distinctive appearance, and health issues including heart defects.[202] Deletion or duplication occur when a region is either lost or repeated. Cri du chat syndrome, which causes the affected individuals to have developmental delays, distinct facial features, and a cat-like cry in infancy. These consequences are a result of deletion on chromosome 5,[203] while some cancers are a result of duplication.[204] Finally, translocation refers to when a segment of one chromosome is broken off and reattached to the wrong chromosome.[205] One form of leukemia is caused by a translocation.

Another type of mutation is a copy number variation (CNV), which may also be considered a chromosomal mutation. CNVs occur when a segment of DNA is either repeated more or less frequently than expected.[206] The two types of CNV essentially differ in the location in which they are found. Gene amplification refers to mutations in which a segment of coding DNA is repeated more times than normal, which is the cause for some kinds of breast cancer.[207] An example of a common copy number variation is the AMY1 gene, which codes for salivary amylase, the enzyme that breaks down starch in the mouth. In populations that consume higher starch diets, individuals tend to have more AMY1 copies, and produce more salivary amylase.[208]

Expanding trinucleotide repeats occur when a repetitive section of DNA is longer than it should be.[209] These repetitive segments are normal and necessary in DNA, but when they are too long, they can cause sever-

---

201 Clancy, 2008
202 U.S. Department of Health and Human Services, n.d.
203 Genome.gov, n.d.
204 Clancy, 2008
205 Morris, 2019
206 Collins, n.d.
207 Clancy, 2008
208 Perry, 2007
209 Clancy, 2008

al kinds of diseases and neurological disorders. Huntington's disease, which typically presents at middle-age, causing cognitive impairment and mood disorders, is caused by an increase of the number of CAG repeats in the HD gene.[210] 26 copies and under is considered normal, while 40 copies and over will cause the disease. Individuals with intermediate numbers will not have the disease, but their descendants are at an increased risk, as less expansion has to occur from parent DNA than a normal individual.[211]

Mutation Causes and Prevention

As previously explained, mutations are very rarely a good thing. Our bodies are highly regulated systems that thrive on consistency in our DNA. Due to this, DNA replication and cell division have many processes to prevent mutations from occurring. When this fails, there are other opportunities to prevent the mutated cell from dividing. Ultimately, however, it is too costly to prevent all changes, and this is how mutations slip through. In addition to understanding what causes mutations, it is useful to take a deeper look into how they are prevented.

Prior to the repair mechanisms, there are many possible ways for mutations to occur. The most basic explanation is that sometimes, errors just happen. At the rate at which DNA replication is taking place, nucleotides can occasionally be missed or mismatched. This is particularly likely in mutation hotspots, which tend to be highly repetitive sequences.[212] In addition, hotspots- and anywhere along a strand of DNA- are vulnerable to mutagens.

Mutagens, put plainly, are things that cause mutations.[213] These include certain chemicals found in the environment, such as arsenic and mercury found in groundwater in many Asian regions.[214] Radiation, such as UV radiation from the sun, is also a mutagen. Large chromosomal deletions and duplications, causing entire chromosomes to be doubled

---

210 Myers, 2004
211 Myers, 2004
212 Clancy, 2008
213 Clancy, 2008
214 Yagi, 2017

or missing, are a result of nondisjunction.[215] During meiosis, it is possible for chromosomes to not separate properly into daughter cells, resulting in some cells with extra chromosomes and others with too few. Intersex conditions, for example, are caused by extra or missing X or Y chromosomes.

An especially relevant cause of mutation is age. As we get older, and our cells undergo more divisions, they become more susceptible to error. This causes several types of disease later in life. The age at which humans have children is also increasing, causing more mutations during fetal development. Down syndrome, caused by the failure of chromosomes to properly distribute during meiosis, is far more likely to occur when a mother is 40 than when a mother is 20.[216]

During cell division, there are several checkpoints to check for errors. If errors are found, division will be paused to allow for repairs. One checkpoint confirms there is no damaged DNA, before it is replicated. The next major checkpoint confirms that all DNA has been replicated, before the cell begins to divide. Finally, the last major checkpoint confirms that all chromosomes are attached to the spindle, which is what pulls chromosomes equally into daughter cells.[217] Once errors are found, how are they repaired?

Each type of error has a specific mechanism to repair it. DNA polymerase proofreads and corrects any mismatched nucleotides, though it still misses many.[218] There is a second chance mechanism that catches more of the errors, though the exact process varies by organism. Nucleotide excision repair is a similar mechanism, but it is able to correct larger strands, rather than a single nucleotide at a time.[219] Base excision repair corrects abnormal and damaged bases.[220] Despite all of these processes, some errors slip through and are able to become permanent

---

215 Morris, 2019
216 Centers for Disease Control and Prevention, 202
217 Morris, 2019
218 Morris, 2019
219 Morris, 2019
220 Morris, 2019

changes in the daughter cells. To a certain extent, efficiency is prioritized over perfection when it comes to cell division. This ultimately gives rise to mutations.

How Mutations Impact Genetics

Broadly speaking, there are two types of cells in the human body: somatic cells and germ cells. Germ cells are those that are used to sexually reproduce - sperm cells and ova. Somatic cells consist of every other type of body cell. Only mutations that occur in germ cells, referred to as germline mutations, can be passed down to offspring.[221] Due to this, germline mutations will be the focus of this discussion surrounding genetics and evolution.

The first thing to consider is how a mutation impacts an individual's fitness. A negative mutation, one that causes premature death or otherwise lowers fitness, will typically only last in a population for a few generations, if at all.[222] Generally the more negative the impact is, the less likely a mutation is to survive in a population. Neutral mutations, with little to no impact on fitness, will not be selected for or against by natural selection. They will persist more or less at random. Mutations that increase an individual's fitness, however, tend to increase in a population over time.[223]

It is also important to consider how a mutation's helpfulness or harmfulness may depend on the environment. For example, the mutation that causes sickle cell anemia persists in many parts of the world, despite having major deleterious effects. This is because a single copy of the mutated allele does not cause symptoms of sickle-cell anemia. The impact that it does have is in fact quite beneficial; it causes resistance to malaria. In parts of the world where malaria is endemic, the mutated allele is useful to maintain.[224] However, outside these geographic locations, the cost of an individual possibly having sickle-cell anemia outweighs the potential benefit.

---

221 Loewe, 2008
222 Loewe, 2008
223 Loewe, 2008
224 Morris, 2019

Perceptions of Mutation in Popular Culture and History

Over many years, particularly since World War 2, mutations have been a popular trope in the media. The threat of nuclear warfare or other disasters brought the concept of mutants into the limelight. This is especially evident in the comics of the last hundred years. Whether it be heroes, such as the X-Men, or their enemies, such as the Brotherhood of Evil Mutants, mutations were perceived as major changes to the human form.[225] Despite heroes using their mutations for good, they were seen as majorly othering forces.

The opposition of good and evil was perhaps a way to cope with the fears of major environmental changes. But why were most conceptions of mutation so scary and large? Before genomic sequencing became possible, not much was known about neutral mutations. As scientists couldn't investigate on the DNA level, only mutations that had an obvious phenotypic impact were understood.[226] Thus, the word mutant was only able to invoke images of barely-human creatures, with supernatural abilities or terrible curses.

The other, far less entertaining side of the historical perception of mutation is fuel for eugenic ideologies. As mutations were seen as largely negative, it was argued that non-European immigrants, people living in poverty, or any other difference that made individuals be viewed as less than, was carried in a mutation of their genes.[227] This led to some unfortunate political movements to discourage reproduction from these groups.

Negative understandings of mutations also contribute to the fear of intentional genetic modification. Ironically, as scientific knowledge of genetics and mutations trickle down to the general population, distrust seems to increase. People wonder what changes are unintentionally being made to the genetic sequence of vegetables when we target specific areas of DNA. Concerns vary from ingesting something harmful in a

---

225 Zehr, 2014
226 Resta, 2017
227 Resta, 2017

previously safe food, to accidentally creating a mutant plant species that will somehow take over the world.[228]

Conclusion

For many years, including in parts of popular culture today, mutations have been perceived as scary, major changes to lifeforms. However, mutations can in reality range from life-threatening diseases, to a change in hair texture, to life-saving alterations. On the genetic level, a mutation could refer to a single missing nucleotide or it could refer to an entire mismatched section of a chromosome. In reality, mutations are most often not scary at all, but rather give rise to the genetic variation that results in the diversity we see all around the globe.

---

228 Resta, 2017

8

What is Cloning?

Written By: Mical Habtemikael

Introduction

Although the term has various vague colloquial usages, cloning is defined as the process of creating an identical copy of a cell or organism genetically.[229] At this point in modern science, everything from genes, cells, tissues, and even whole animals are being cloned. However, the process of cloning has long since existed in nature and occurs all around us at any given time. Bacteria are a good example of this because they make copies of themselves and reproduce. In humans, twins are a great example of the existence of cloning genes. Identical twins are a result of a fertilized egg splitting in two.[230]

Amongst these examples, there are a number of different types of cloning processes in many different areas of life with their respective uses. Generally, artificial cloning can encompass quite a number of subset cloning groups. Take for instance, the process of molecular cloning, which refers to the reproduction of large quantities of deoxyribonucleic

229 Rugnetta, 2020
230 National Geographic Society, 2019

acid (DNA) or reproductive cloning which uses technology to produce animals who share genetic characteristics with one another and therapeutic cloning, which produces a number of cloned embryos.

Background

The origin of reproductive cloning was a result of attempts to replicate sea urchins in the 1890s when Hans Dreisch attempted to isolate sea urchin embryos and observed how they developed into larvae. Following this experiment, in 1902, Han Spemann used the same procedure to split and isolate cells from salamander embryos. A year later, the US Department of Agriculture coined the term "clon" to refer to the aesexual production of cells or organisms and this term was eventually changed to what is now "clone." The root of the word derives from the Greek klon, which means "slip" or "twig." This term has been used by Horticulturists for over a century to refer to nature's replication of its cells. In fact, plant cloning can be traced as far back as 1838 to German scientists Matthias Schleiden and Theodor Schwann who presented their cell theory. This theory presented the idea that all cells arise from preexisting cells. [231] And thus this theory gave birth to the idea that plants could be regenerated from an isolated cell. Closing in on the mid 1900s, Spemann's experiments eventually graduated to transferral of the salamander embryos' into one another with success. This trail of experiments being tested in frogs, extended to fish and eventually larger mammals.

Molecular Cloning

Molecular cloning is a branch of artificial cloning that uses a fragmented piece of DNA of a model organism, referred to as a restriction endonuclease (REs) which is inserted into another piece of DNA called a cloning vector. Cloning vectors are made up of plasmids, bacterial viruses or yeast artificial chromosomes that take the shape of a small circle. The process takes place when the genome fragments, a set of genetic material, is moved and populates the model organism or host cell and is referred to as the gene library. The process of gene isolation is done by finding fragmented DNA or RNA samples that come from the genes of closely related organisms and under a probe encompass a DNA

---

231 Owen, 2019

sequence that is complementary to that of the isolated gene.

Reproductive Cloning

Reproductive cloning is most often linked with the process of replicating animal genes. This process isolates mature eggs from female animal species and pierces them with a microscopic narrow glass tube to vacuum the egg nucleus out.[232] The procedure of removing the nucleus from a cell is referred to as enucleation. This now enucleated egg is then fused to the body of the animal intended to be cloned and activated with either chemicals or an electric current. After the chemical or electrical activation, somatic cell nuclear transplantation (SCNT), the egg divides itself and experiences growth into a newly formed embryo. Once the embryo is formed, it needs to be placed within or in the proximity of the womb of a species-sharing female animal in order for it to receive the hormones necessary to grow beyond that point. Eventually the pseudo-pregnant animal will have to give live birth to the clone.[233]

Therapeutic Cloning

Much like the reproductive cloning process, therapeutic cloning involves embryos that have undergone the somatic cell nuclear transplantation (SCNT) and then are disassembled in a laboratory.[234] The fragmented embryo is then gestated into the surrogate mother but only until the fetal stage at which point it is aborted and the cells from the fetus are then used to concoct a cell culture.[235] The culture of these specific cells are then used to make embryonic stem cell (ESC) cultures which are pluripotent meaning that they have the capacity of replicating different types of cells.

Gene Therapy

The benefits of molecular cloning are evident in a lot of today's pharmaceuticals, and they include human insulin, growth hormones, fertility drugs and many vaccines. Molecular cloning is also used to screen

---

232 Buratovich, 2018
233 Boylan, 2002
234 Buratovich, 2018
235 Buratovich, 2018

humans for genetic conditions.[236] Pharmacogenetics or the use of cloned genes to screen patients can also predict the efficacy and toxicity of many of the drugs that that patient might be a candidate for.[237] Cloned genes can also be used to treat patients who suffer from deficiencies in their genes that may contribute to their development of cancer, immune system deficiencies and other blood based defects. Using molecular cloning to deliver cloned genes to the patients who suffer from genetic diseases is known as gene therapy.

Animal Cloning

When animal cloning is mentioned, it is usually in regard to higher vertebrates and mammals. Using a differentiated cell and making a copy of it. Despite cells having specific function within the body, the nucleus houses the genetic data that is supposed to help support the growth of the organism. To successfully clone an animal the data taken from the cell should be reformed to an undifferentiated cell.

Despite the purpose of cloning being to have a replica of a biological organism, it is not possible to have a clone that is exactly identical.[238] There are some differences which are a result of environmental factors and mitochondrial factors. The mitochondria, here, is not to be confused with the nuclear genome, which is an exact replica. The mitochondria can influence the biological organism's phenotype. The phenotype is an organism's features that show themselves. Phenotypes can include but are not limited to eye colour, hair colour, hair texture, skin colour, size and more.[239]

The first successful documentation of successful animal cloning was 1996.[240] The animal that was cloned was a sheep and her name was Dolly. There were 277 recorded attempts of cloning in this particular instance. Dolly was the only one of the attempts that came out alive. After this precedent was set, it was quickly followed by more successful instances of animal cloning. Animals including mice and cattle were suc-

---

236 Uddin et al., 2020
237 Cibelli, 2002
238 Robinson, 2019
239 Houdebine, 2003
240 Robinson, 2019

cessfully cloned.[241] Although mice were successfully able to be cloned, the scientist involved found that they had trouble reaching that success, even more so than the cattle. After being successful with the cloning of mice and cattle, scientists moved on to buffalos, cats and pigs.

Some of the problems scientists have run into surrounding animal cloning is how it is not efficient. Around the mid 2010's an experiment done with the hopes of successfully cloning animals was only successful at a rate of 10%.[242] It is important to note that animals that are a product of animal cloning are usually larger than their counterparts that were reproduced naturally. Due to their unnatural size, they usually cannot be delivered naturally and are delivered through a C-section. This has a negative effect on their chances of survival and that is why they have a higher morbidity rate. In addition to a riskier delivery, animals that are a product of animal cloning tend to have shorter lives. Signs of aging usually begin earlier in cloned animals. The first successfully cloned animal Dolly the sheep only lived for roughly 50% of the span as her naturally created counterparts.

Although, there are definitely some downsides to animal cloning, there are also added benefits. Some animals are high sought out and if scientists have the ability to successfully clone them this would-be a plus. Many humans use and depend on livestock to make a living and as a part of their diet. Animal cloning could potentially help make securing livestock easier. If possible, animal cloning would be ideal in regard to endangered animals. This could help vulnerable species re-populate and help balance their ecosystem.

There is still much to learn about animal cloning; both China and the United States have not discovered discrepancies between healthy cloned animals and naturally produced animals. In 2015 the European Union's parliament implemented a ban that would not allow for cloning to be done on farm animals. Part of the arguments against animal cloning was based on concerns for the animals and their well-being. Since animal cloning is not efficient, many animals end up dying in the process. There was also concern surrounding the quality of the animal meant

241 Robinson, 2019
242 Jabr, 2013

for human consumption and the impact that it could potentially have.

Horticulture Cloning

In plants, natural cell replication exists in two types of organisms, pro-karyotic and eukaryotic organisms. Prokaryotic organisms are micro-organisms that are lacking a cell nucleus such as bacteria and are able to replicate themselves through a process of binary fission or budding.[243] A eukaryotic organism by comparison is an organism possessing a cell nucleus - this includes human cells who are put through the process of mitosis.The greatest impact of the isolation of plant cells, tissue and or-gan cultures are the ability for humans to more rapidly produce them. Humans now require less space and plants can be produced year-round [244] This is done by propagating vegetable crops, fruit crops, floricul-ture species and even woody species. The cloning of different plant life requires significant political, economical and environmental consider-ations, especially for countries that are able to implement these practic-es.

To embark on the process of horticulture cloning, one must make two significant decisions of the type of propagation technologies they are to employ.[245] The first of which is the cultivation of a whole field of genet-ically identical plants otherwise known as a monoculture which leaves the entire crop available to disease or pests. The other consideration is the generation of populations of clones through tissue culture technolo-gies and incurring the potential problem of a genetic abnormality that would then be present in a whole population of plants produced, which is termed somaclonal variation.[246]

In agriculture, the continually rapid expansion of the human popula-tion has required the increased production of the global food supply. This increased demand has promoted genetic modification techniques. Cloned plant genes are used to generate transgenic food crops, meaning they contain genetic material with artificially introduced DNA. Crops

---

243 Rugnetta, 2020
244 Watts, 2020
245 Owen, 2019
246 Owen, 2019

enhanced through molecular cloning intervention display more advantageous traits such as a reduced dependence on agrochemical applications such as herbicide, a display in increased nutritional value and are able to withstand extreme temperatures.[247]

Genetically modified organisms (GMOs) in agricultural products while questioned for their ethical merit have been planted by 18 million farmers in 27 countries worldwide. In fact, 87 percent of global genetically modified crops that are grown worldwide are produced in developed countries; however, developing nations are slowly enhancing their crops with crucial nutrients. For instance, countries in Asia that produce rice have seen GMO enhancement for iron and other vitamins and have aided in the alleviation of malnutrition. Recombinant gene technology has also made it possible for plants to survive extreme temperatures and convert atmospheric nitrogen into a more usable form of nitrogen. Additionally, genetically modified plants are able to produce their own resistance to pests and pathogens which eliminates the need for chemical pesticides.[248] An example of this is the production of Roundup Ready soybeans that were engineered to resist herbicide used on weeds commonly found where soybeans grow. Similarly Bt corn contains a bacterial gene that increases pest resistance.[249]

Existing studies conducted by the US Department of Agriculture (USDA) have closely monitored genetically modified crops to ensure that they are not as disruptive to the environment. This has included thousands of completed and pending field trials for various crop species that include but are not limited to potatoes, cotton, canola, cucumbers, etc.[250] However, in the realm of public perception, consumers do not seem to have had their concerns alleviated when it comes to the existence of GMOs.

Conclusion
In summary, while cloning has existed for quite some time there are

247 Watts, 2020
248 Thieman, 2013
249 Owen, 2019
250 Weasel, 2009

still a lot of political, economical and environmental considerations that need to be made moving forward. The world has also yet to assess the full impact of the ways our food and medical systems have been intervened by cloning technology.[251] Not to mention the significant amount of controversy that surrounds the field of cloning. Human reproductive cloning for one has remained universally condemned due to the psychological, social and physiological risks associated with cloning and the estimation of a high number of deaths. Additionally there is a fear that human cloning will fan the flames of eugenics ideals which promote the once popular notion that the human race could be improved upon.[252] Many also consider the use of therapeutic cloning an issue of ethics bause of the manufacturing and subsequent destruction of human life to perform its medicinal tasks. But with all this in mind there is one thing that can be said with certainty. and that is that human advancement is owed in large part to the power of cloning.

251 Rugnetta, 202
252 Rugnetta, 202

How Does Genetics Impact Us?

Written By: Mackenzie Schuler

How do Genetics Impact Health?

Genetics have been linked to health problems in many situations. Whenever someone goes to the doctor, the doctor asks for a family medical history. This is so they can determine if that person may have a predisposition for mental, physical, or emotional diseases.[253] Predisposition is defined as the likely risk of experiencing or suffering from a particular disease.[254] If someone has a predisposition to a certain disease, this does not mean they will certainly have or experience the symptoms of the illness; it just means that there may be an associated risk.[255] Some of the questions doctors ask about family medical history are relatively simple: ethnicity, mental health conditions, pregnancy complications. However, some of the questions are hard to answer as they may require information about deceased family members. Due to this, some people prefer to prepare a concise medical history prior to a doctor's visit. It's important to provide doctors with medical records so that they have all the appro-

---

253 Mayo Clinic Staff, 2019
254 "Predisposition," n.d.
255 Mayo Clinic Staff, 2019

priate information to help and support their patients.[256]

Biologically, every individual inherits half of their genetic profile from each of their parents.[257] Certain genes inherited can represent possible concern for future medical conditions, but certain genes may also represent a lack of concern for medical conditions. An example of this are the BRCA genes. The BRCA genes are also known as "the BReast CAncer gene". Every person has both the BRCA1 and BRCA2 genes, but not every person experiences the same representation of the gene. Although these genes themselves do not cause cancer, there is evidence suggesting that they may actually prevent breast cancer and tumour growth on the breast tissue by repairing DNA tears that can be associated with cancer or tumours. BRCA genes are known as tumour suppressing genes due to this evidence. However, in some cases, there are BRCA mutations that do just the opposite: they don't suppress the tumours, therefore the BRCA gene isn't working properly.

A very small portion of the population (1 in 400) carry a mutated form of the BRCA1 or BRCA2 gene. The mutated form, as mentioned earlier, may be ineffective when it comes to preventing breast cancer. A considerably large portion of women will experience breast cancer in their lifetime (1 in 8). The BRCA gene has been associated with an increased risk of developing breast cancer, and has been associated with experiencing a cancer recurrence.[258] A cancer recurrence is a second cancer that has returned, and can present as the same original form of cancer or as a secondary form.[259] Breast cancer itself is a very treatable form of cancer if it is treated in it's very early stages (Stage 1 breast cancer has a 5-year relative survival rate of 100%).[260] However, approximately 10% of women with breast cancer have a BRCA mutation, suggesting that there is a link but it may not be a significant link.[261]

---

256 Mehta, n.d.
257 Mayo Clinic Staff, 2019
258 BRCA: The Breast Cancer Gene, n.d.
259 "Recurrent Cancer," n.d.
260 Stages 0 & 1, n.d.
261 BRCA: The Breast Cancer Gene, n.d.

There are three common types and three rare types of sickle cell disease. The three most common types are HbSS (sickle cell anemia, which is the most severe form of the disease), HbSC (which involves the odd hemoglobin "C"), and HbS beta thalassemia (which is a different type of anemia compared to HbSS). The three most rare types are HbSD, HbSE, and HbSO.[262] People with sickle cell disease have a considerably short life expectancy, as half are expected to live a little bit beyond 50 years of age. There are treatments available, with the most commonly used treatments being antibiotics for infections and blood transfusions. However, the only cure for sickle cell disease is bone marrow transplants.[263] There are limitations to this cure; there is about a 25%-30% chance of finding a compatible bone marrow donor, regardless of how many siblings the patient may have.[264]

Although some of the associations between genetics and health are related to specific diseases, there is now research being done to investigate if genetics plays a role in how people respond to drug treatments for multiple diseases. It's well-known that everyone will likely have a different reaction to certain pharmaceuticals, but there is a possible link between genetic factors and how people respond to pharmaceuticals. There is a possibility that genetics impact the individual's sensitivity and metabolic response to a drug.[265]

Mostly every gene has been linked to some form of genetic condition. About 6 in every 10 people will be personally affected by a genetic condition, with a range in severity. Many health conditions run in families, suggesting a genetic connection.[266] For example, cataracts are considered a non-genetically inherited condition. Cataracts are an eye condition that causes cloudy vision. Cataracts develop over time due to aging and are typically treated using surgery. However, there are some genetic conditions that increase the risk of developing cataracts such as a condition called congenital cataracts, which are cataracts a person is born

---

262 What Is Sickle Cell Disease?, 2020
263 About Sickle Cell Disease, 2020
264 Grisham, 2019
265 Hernandez & Blazer, 2006
266 Department of Health, Government of Western Australia, n.d.

with or develops during childhood or adolescence. Congenital cataracts may be genetic or may be caused by certain conditions such as rubella.[267]

Although someone may have a predisposition for a certain health condition, there are several other factors contributing to the risk of having the disease. Some of these factors are environmental and lifestyle choices. An individual's genetic code can not be modified, but there are still some protective factors that can possibly reduce the risk of disease.[268] For example, eating healthy meals and getting blood pressure checks regularly may reduce the risk of cardiovascular and heart diseases.[269]

What are Genetic Mutations?

A genetic mutation refers to a change in the sequence of a gene. Mutations are generally associated with an increased risk of certain diseases, disorders, syndromes, or illnesses.[270] There are four major causes of genetic mutations: infections, exposure to mutagens, errors in cell division, or exposure to radiation. There are six general forms or types of genetic mutations: deletion, insertion, substitution, duplication, inversion, and translocation.[271] Genetic mutations are permanent modifications that modify the DNA sequence of a gene. Inherited mutations are hereditary passed down from the parent to the child. Non-inherited mutations are not hereditary and occur as a result of an event or exposure during someone's lifetime.[272]

How do Genetics Impact Personality?

There is no singular gene that determines personality. Instead, there is an array of genes that coordinate together to present a personality. However, there is an active argument over whether nature or nurture has a larger impact on the presentation and development of personality.

---

267 Mayo Clinic Staff, 2018
268 What Does It Mean to Have a Genetic Predisposition to a Disease?, n.d.
269 How to Prevent Heart Disease, n.d.
270 Anzilotti, 2021
271 Collins, n.d.
272 What Is a Gene Variant and How Do Variants Occur?, 2021

Luckily, there is an entire field devoted to the scientific study of gene association and personality traits: molecular genetics.[273]

Children begin to develop a personality at a young age. Parents tend to study their children and try to predict the kind of person their child will grow up to be.[274] Temperament is the part of a child's character that affects behaviour and mood development.[275] Temperament is biologically based, beginning from birth to shape a child's personality. It works in tune with external factors, such as the environment, to contribute to the growth and development of a child's personality. As children grow older, their temperamental characteristics modify and adjust to match their new biological maturity. As children get better at self-control, self-regulation changes become more apparent in the personality of the children.[276]

However, there are other qualities that impact the development of a child aside from temperament. Motivations, socialization, values, coping mechanisms, and maturity are just some of the many factors that go into personality. Temperament works alongside these factors and other external factors to build a personality. Temperament is the biological, or genetic, part of personality while other factors, such as socialization, are the environmental part of personality.[277] There have been multiple twin studies to determine if personality is a perfectly 50%-50% split in heredity. Genetics make a sufficient impact on observable characteristics, such as personality. Twin studies have been quite consistent, and the findings suggest that personality traits are a result of genetic factors and non-shared environments as opposed to shared environments.[278]

Personality is a trait determined by many factors, and as such there are many genes that can be held accountable for personality. As it's known,

---

273 Walinga & Stangor, n.d.
274 Thompson, 2021
275 "Temperament," n.d.
276 Thompson, 2021
277 Thompson, 2021
278 Kreuger et al., 2008

genetics play a role, but there is no singular gene that is responsible for whether someone is social or whether someone prefers to spend more time alone (extroversion vs. introversion). The environment does play a considerable role, as people are socialized and brought up differently depending on where they live and other environmental factors. There are five traits that have been linked to heritability, according to molecular geneticist Dana Bressette: extraversion, agreeableness, openness, conscientiousness, and neuroticism. These traits tend to coordinate with environmental factors (such as education and nurturing) to allow for the full development of a unique personality for every individual.[279]

Although anxiety is not a personality trait, many individuals experience troublesome levels of anxiety and genetics has a role in this. There is no singular gene that determines whether a person will struggle with anxiety, but there are several genes that work together and present a predisposition to anxiety. The tendency to experience anxiety may be passed down genetically through these genes.[280]

Research in honey bees suggests that certain components of RNA may be associated with predicting behaviour. This may be evident due to the idea that DNA and RNA also work together to produce certain hormones and neuropeptides. Hormones play a role in actions and behaviour, which is a very important part of personality. Genes quite clearly may have a contribution in personality, but there is not enough scientific evidence at the present time to clearly define how big of an impact genetics has on personality. However, there is promise of future studies providing more information for this field.

How do Genetics Impact Child Development?

Genes effectively begin child development. The development of a child starts prenatally (in the uterus), with the presence of the ovum (commonly called the egg) and sperm. Both the sperm and the ovum contain a genetic code consisting of chromosomes that combine to make up the 46 chromosomes of the human body. The ovum has 23 chromosomes

---

279 Sokol, 2021
280 Sokol, 2021

and the sperm does as well.[281]

There are some cases of sex chromosomal abnormalities that affect child development.[282] In Klinefelter's Syndrome, a boy has an extra X chromosome, leading to lower levels of testosterone and to problems with testicular growth, lack of muscle mass, minimal body and facial hair, and in certain cases, less production of sperm. In addition to that, there are development issues as well, such as delays in speaking and delayed motor development.[283] Turner Syndrome is another example of a sex chromosomal abnormality. Turner Syndrome only affects women and is a result of when one of the X chromosomes is missing. This often results in an array of developmental issues, such as short height, inability to develop ovaries, and cardiac defects.[284]

Trisomy 21, known as Down Syndrome, is the most common chromosomal disorder.[285] In Down Syndrome, the individual has an extra chromosome (three chromosomes instead of two) at site 21. This affects the way that the child will develop, both mentally and physically. Typically, people with Down Syndrome have an IQ in the relatively low range, and experience delayed speech. Down Syndrome tends to occur in approximately 1 out of 700 babies. The exact causes for Down Syndrome are still unclear, and it is also relatively unclear how to determine whether someone will be born with Down Syndrome. However, one known factor that increases the risk of having a baby with Down Syndrome is the mother's age. If the mother is over the age of 35, they are more likely to give birth to a baby with Down Syndrome than a mother younger than the age of 35.[286]

How do Genetics Impact our Appearance?

As mentioned in the section above, people tend to have 46 chromosomes, unless they experience a chromosomal abnormality. These genes

---

281 Cherry & Block, 2020
282 Cherry & Block, 2020
283 Mayo Clinic Staff, 2019b
284 Mayo Clinic Staff, 2017
285 Cherry & Block, 2020
286 What Is Down Syndrome?, 2021

makeup a person's genotype. The way they physically present themselves is called the phenotype. As discussed in earlier chapters, there are homozygous (the same) and heterozygous (different) types of gene pairs. In addition, there are recessive and dominant traits.[287] When certain genes combine, they form either a homozygous or heterozygous pair, but how the phenotype presents itself depends on whether the traits are dominant or recessive.

Oculotaneous albinism is an example of a physical appearance that is affected by genetics. This condition impacts the pigmentation of hair, skin, and often eyes. There are four types of albinism, from Type 1 to Type 4. People with type 1 tend to have white hair, extremely pale skin, and light-coloured irises. Type 2 is similar to type 1, except that the skin has more of a cream colour and the hair may be light blond, blond, or even light brown. Type 3 usually affects dark skinned people and is known as rufous oculocutaneous albinism. The skin pigment tends to be red-brown, the hair presents as a ginger or red colour, and the eye colour is typically hazel or brown. Type 4 is similar to type 2. Albinism is inherited through an autosomal recessive genetic pattern, meaning both copies of the genes have mutations. Typically, parents do not show signs of albinism even though their children do.[288]

Eye colour is another example of a genetic trait that affects appearance. Eye colour is simply the pigmentation of the part of the eye known as the iris. Eye colour is determined by genetics. Eye colour is associated with melanin, which is a pigment. People with brown eyes or dark eyes tend to have more melanin in the iris, while people with lighter eye colours (such as blue) tend to have less melanin in the iris.[289] There are two main genes that play a role in determining eye colour, and both genes are in one region of chromosome 15. The gene OCA2 codes for a protein that affects the quality and quantity of melanin in the iris. The other gene is HERC2, which is responsible for controlling the expression of OCA2. The inheritance pattern of eye colour is not easily predictable. Although most of the time you can guess what colour eyes a child will have by

287 DNA DETERMINES YOUR APPEARANCE!, n.d
288 Oculocutaneous Albinism, 2020
289 Is Eye Color Determined by Genetics?, 2021

looking at their parents, it is not a definitive way to determine eye colour due to variants that may occur. The allele (variant of a gene) for brown eyes is the most dominant, while the allele for blue eyes is the most recessive, resulting in the allele for green eyes to be recessive when paired against a brown eye allele, but dominant over the blue eye allele.[290]

Hair colour is also determined by genetics. As mentioned earlier in the discussion regarding albinism, melanin helps determine hair colours. There are two types of melanin: eumelanin, which results in darker hair colours, and pheomelanin, which results in lighter hair colours. The most well-studied gene thought to determine hair colour is MC1R. This gene instructs the production of the protein melanocortin 1 receptor. When the receptor is activated, eumelanin is produced. When the receptor is blocked, pheomelanin is produced.[291]

Once again, the phenotype for hair colour depends on the dominance and recessiveness of the alleles. For example, brown hair is dominant while blonde hair is recessive. This means that if one parent had brown hair and the other had blonde hair, they are more likely to have a child with brown hair, depending on whether the brown hair parents genetic code is homozygous or heterozygous. However, hair colour is not always determined solely by genetics. Environmental factors play a role which results in changing hair colours. Smoking or spending a lot of time in the sunlight can cause a hair colour to lighten or fade.[292]

There are other physical traits impacted by genetics such as dimples, ears, and widows peaks. The genetic code given to people by their parents impacts most things in their lives, especially their physical appearance. There are many different gene variations and determinants that affect how a gene will represent itself, and there is more research being done to determine which genes specifically play a role in the presentation of physical appearances.

---

290 Robertson & Anderton, 2019
291 Is Hair Color Determined by Genetics?, 2020
292 Riesen, 2019

How do Genetics Impact Behaviour?

Instincts are inborn tendencies that correlate to behaviour. Instincts require no thought and are considerably difficult to modify through learning. There is some role of genetics in behaviour, as instincts are a form of behaviour and they are an inborn tendency, but there is not enough scientific evidence and there is too much debate in the field to truly determine how big of a role genetics play.[293] However, the field of behavioural genetics is an expanding field and there are new studies being done to provide more information on the link between genetics and behaviour.[294]

The way that we can further study the influence of genes is through heritability estimates. A heritability estimate is a mathematical formula that can suggest how impactful certain genes will be on behaviour.[295] Although there is not much known about the link between genes and behaviour, there is some evidence suggesting that genetics impact learning and there are certain behaviours that are learned.[296] Examples of learned behaviours are cooking, sports, and playing instruments.[297]

The Benefits of Genetics

Genetics have many benefits, however in some circumstances, as mentioned earlier in the chapter, people may be subject to experiencing genetic disorders. These disorders can be dangerous if they are untreated and go undetected. Some examples of these disorders are Cystic Fibrosis, Sickle Cell Anemia, and Huntington's Disease. Genetic testing, or genotyping, can help determine how individuals will experience predispositions for disorders. However, a predisposition does not necessarily mean that someone will experience a disorder. Genetic testing can help individuals with complex family medical histories and help them make decisions for the future.[298]

---

293 Breed & Sanchez, 2010
294 Genetics and Behavior, n.d.
295 Asendorpf, 2001
296 Genetics and Behavior, n.d.
297 Learned Behaviors In Humans, 2019
298 The Benefits of Genetics, n.d.

What is Genotyping?

Genotyping is genetic testing that can determine predisposition to genetic diseases and disorders. This testing process compares DNA samples to one another to compare variations and determine the complexity of DNA strands. Genotyping can help determine risks through these comparisons.[299]

Genetics and Us

Many people have predispositions to many things. Although genes play a role, environmental factors must be considered when determining how genetics impact people. Genetics impact appearance, behaviour, personality, and health. Genotyping is a common way to determine genetic codes and to determine risks of medical conditions. Genetics have many benefits and numerous impacts.

---

299 What Is Genotyping?, n.d.

Evolution in the Modern World

Written By: Mya George

Introduction

Charles Darwin first proposed evolution in 1859, and since then, the understanding of and perspectives on the theory have themselves evolved.[300] Evolution has frequently been presented in contradiction to religious creationist beliefs, and while some Christian denominations strictly oppose the theory, many religious groups have come to accept evolution alongside biblical accounts of creation. In popular media, evolution is utilized for drama and plot in popular science fiction and consequent adaptations such as with the X-Men or Ray Bradbury's A Sound of Thunder. Confusion and misunderstandings have stemmed from the creative license used with science in this type of content. Furthermore, the way evolution is taught in Canadian classrooms also shapes the public understanding of this concept. It is commonly introduced later on in secondary education and within elective upper-level biology courses; the students most vulnerable to misinformation are those who are less likely to receive instruction on the subject of evolution. The understanding of evolution in the modern world is affected by many factors, and

---

300 Chapter 2

will continue to be shaped by influential aspects of people's lifestyles.

Religious Beliefs and Evolution

Evolution and other scientific concepts are commonly seen as contradictory to religious beliefs, especially with regards to Christianity in North America. In Genesis, the first chapter of the Old Testament, it is told that God created the heavens and the earth, and all that inhabits and characterizes the earth, over the course of 6 days. Following His creation, He came to rest on the seventh day which was called the Sabbath. A literal interpretation of this creation story is not supported by the concept of evolution first proposed by Charles Darwin. Furthermore, in Genesis it says "God created humankind in his image... male and female he created them".[301] Some Christian denominations reject the idea of human evolution because they believe people were made in God's own likeness and through His will, a concept that is known as intelligent design. Charles Hodge (1797-1878), the Presbyterian theologian wrote in his 1874 critical article "What is Darwinism?" that "the denial of design in nature is virtually the denial of God".[302] Intelligent design serves to credit an omniscient creator for all the development on earth, instead of attributing it to evolution over time. Conversely, another popular belief surrounding human evolution is that God can work through intermediate processes to create life. This concept is known today as evolutionary creationism which centres on scientific processes at play originating from a higher power.

In 1950, Pope Pius XII argued in his writings on the human race that evolution was compatible with the faith as long as one understood that God Himself is responsible for the creation of the human soul.[303] The Catholic Church today is in support of evolutionary creationism, although the Catechism (the book containing Church doctrine) states that Catholics may believe in either the creation story as written in Genesis or in evolution guided by God. The General Assembly of the United Presbyterian Church came to a similar agreeance in 1982, that evolu-

---

301 New Revised Standard Version Catholic Edition Bible, 1989, Genesis 1:27
302 Series, 2006
303 Ayala, 2020

tion did not conflict with Biblical teachings on creation.[304]

Christian denominations who take a literal understanding of the Bible, such as Christian Fundamentalists, Seventh-day Adventists, and Pentocostals, were not as quick to accept the scientific theory of evolution or to promote it in public education. In addition to reading the books of the Bible as the literal truth, these denominations have also believed that earth and life is a more recent creation. During the 1920s, Biblical Fundamentalists in the United States influenced state legislators to debate anti-evolution legislation. The states of Arkansas, Mississippi, Oklahoma, and Tennessee all banned the teaching of evolution in public schools before the Supreme Court declared this type of legislation to be unconstitutional in 1968. Instead of evolution, Biblical Fundamentalists have pushed for the teaching of creation science.

Creation science encompasses teachings from all throughout Genesis: that all organisms came to be through God, that the world is relatively new and has only existed for a few thousand years, and that the biblical flood killed all but one pair of each animal species (excluding humans, of which Noah and his immediate family all survived). The flood and Noah's Ark is one of the "most recognizable scriptural stories in our cultural repertoire".[305] In 2016, the theme park Ark Encounter that centers around a sizable reconstruction of Noah's Ark, was opened in Kentucky. In addition to recreating the Genesis flood narrative for visitors, this park serves to further teach visitors about creation science. In its second year of operation Ark Encounter drew in 1 million visitors.[306] The success of this creationist theme park, the widespread media attention it has received, and its creation in the first place helps to illustrate the continued support for creation science in communities across North America. Additionally, in many states, creationist Christian minorities have continued advocating for creation science to be taught alongside evolution and in equal parts.

Evolution is not only a divisive topic within branches of Christianity.

---

304 Ayala, 2020
305 Bielo, 2018
306 Knight, 2018

Other world religions also have conflicting views on how the scientific theory of evolution computes with their own teachings and scripture relating to creation. In the Jewish faith, God is said to have created the world with a purpose known by Him.[307] He then created natural laws and tasked humans with a set of religious and moral obligations. Based on this understanding of the purpose of creation within Judaism, evolution does not conflict. Regarding the details of creation, like in Christianity, there is a difference of opinion on whether or not the biblical story should be understood as truth. These differences surround how God acted to create the universe and the amount of time it took. Classic Rabbininic teachings say that the universe was created from nothing by God around 6,000 years ago and that He fashioned the first humans from clay. This understanding of creation is widely accepted by Orthodox Jews today. However, most Conservative and Reform Jews have come to view evolution as scientific fact and this has been the majority since the mid 1900s. Evolution in this context allows for a greater interpretation of Genesis and other scripture relating to creation in Judaism.

Prevalent Anti-Evolution Perspectives

Other arguments against the theory of evolution can stem from outside religious spheres of influence. One popular argument is that evolution would not be capable of producing living beings of increasing complexity over large periods of time because this would be in conflict with the second law of thermodynamics.[308] The second law of thermodynamics states that there is a loss of energy as it is transferred and that order within a system decreases over time. However, this law applies to isolated systems and the earth itself is not a perfect isolated system because there is an energy exchange with elements in space such as the sun. A more sophisticated version of this argument says that although the earth is not an isolated system, the energy of the sun increases disorder to accelerate breakdown and decay on earth.[309] Physicists and astronomers have found this to be untrue mathematically as evolution occurs over millions of years.

---

307 Steinberg, 2010
308 Smith, 2009
309 Oerter, 2006

While intelligent design, the argument that specific details of living systems are proof of an intelligent creator, is a religious anti-evolution perspective, it is closely connected to another argument which exists outside a religious context: irreducible complexity (IR). IR centres on the belief that stochastic processes cannot produce complex systems and biological processes that are found in nature. This argument was outlined by Micheal J. Behe, a professor from Lehigh University, in the 1996 book Darwin's Black Box.[310] Gradualism is the specific Darwinian concept which is argued against in IR, and if biochemical systems did not develop gradually, then this points towards intelligent design. However, Darwin's theory included mentions of how natural selection allowed for gradual changes which allowed for bigger advances in time. Behe's specific example in his book was the complexity of the blood clotting cascade, one bodily system where several intricate parts function in tandem for a very specific function.[311] If any one small detail of this system was not present, it would fail to function. For such a system to come together in such detail, Behe again credits an intelligent creator. The scientific community largely rejects the arguments of irreducible complexity and intelligent design, but there is continued support for these arguments from the general population. While they are in the minority, some proponents of intelligent design such as Behe, are themselves scientists and professors.

Evolution as Seen in Popular Media

Bad science is often used in television, film, and written fiction to create elaborate plots and drama at the expense of the public's conception of scientific concepts. Evolution is affected by this especially with theories being overly simplified, misconstructed, or used as an intelligent tool to bring new species into existence. A popular falsehood across different types of media is that Homo sapiens evolved from monkeys or apes whereas in reality, humans and apes are believed to have evolved from a common ancestor. An opening "couch-gag" from season 18 of The Simpsons had a sequence where the main character, Homer Simpson travels back in time to the Cretaceous period.[312] From there, a sin-

---

310 Draper, 2002
311 Draper, 2002
312 The Simpsons Wiki, n.d.

gle-celled organism made to resemble Homer separates, becoming a jel-lyfish and then a fish that grows legs to walk on land. The ever-evolving Homer is shown trying to survive amongst dinosaurs made to look like other characters in the show before becoming an ape and then finally, a man. While this scene is over-dramatised for comedy, it further propagates the misconception that humans evolved from apes.

Evolution as an "intelligent force" with a greater purpose beyond survival is a frequent plot device in science fiction. A famous example of this is Jack Kirby and Stan Lee's X-Men. The X-Men are presented as an advanced human species with specific mutations resulting in them having superhuman powers.[313] These so-called mutants are part of a fictionalized next stage of human evolution and in the comics are classed as "Homo sapiens superior". The mutations are nonsensical in that each person gains a unique ability, where many of these abilities do not offer a particular survival advantage.[314] Abilities such as death perception, understanding the consequences surrounding one's death, and self-duplication provide no-direct advantage for a person. Conversely, some other abilities seen in the comics seemingly take inspiration from evolutionary traits from other organisms, with mutants having the ability to secrete acid or have bone protrusions forming animal-like claws. Despite all this, it seems that the general trend of abilities and mutations does not manifest in relation to environmental factors. Additionally, this series uses evolution as a magic-like concept. One character, Darwin, was capable of "adaptive evolution" in the comics, changing his body instantaneously to survive in seemingly unsurvivable conditions.[315] While the X-Men are figures of highly imaginative fiction, mentioning evolution and related concepts in connection with fictional elements can alter people's perceptions and understanding.

How Do We Teach Evolution in Canada?

Teaching evolution is important to include in high school curriculum not only as a part of foundational science literacy, but also to educate future voters, workers, and potentially scientists. Having an understand-

---

313 Olsen, 2017
314 Olsen, 2017
315 X-Men Movies Wiki, n.d.

ing of science relating to the natural world and the evidence in support of it, is beneficial for students today who will become future leaders. Online polling conducted in 2019 found that 61% of Canadians believed that human beings did or likely did evolve from less advanced life forms over time.[316] However, it is notable that this percentage decreased 5 points from the poll held a year earlier. This poll did not address or assess people's general knowledge on the theories of evolution, but instead looked at what evolution primarily refers to, how human beings came to be the advanced species existing today. Additional data from this 2019 poll found that 38% of Canadians believed that creationism should be taught in schools, 39% believed that it should not be taught, and the remainder were undecided.[317] While public acceptance of the theory of evolution has increased over time, the public is still split on the role it should have in education. Additionally, much of what the general population knows about evolution is riddled with misunderstanding. A general lack of knowledge and the persistence of pseudoscience can be, in part, contributed to an education system that does not prioritize evolutionary teachings and allows for significant differences regarding how any content is taught.

Canadian institutions have had a number of controversies regarding evolution in education, with anti-evolution groups continuing to promote equal education of creation science alongside the theory of evolution. In the 1990s, the Abbotsford School District in British Columbia (BC) invited creationists to present their beliefs in classroom settings.[318] This practice was halted by the Minister of Education in BC, receiving widespread media attention, and was eventually involved in legal action by the BC Civil Liberties Association. In 2000, the Ontario government was criticized for the new curriculum update that was accused of "[downplaying] evolution education".[319] In the same year, the Home and Schools Federation of Prince Edward Island (PEI) called for the education department to ensure that creation science was discussed in equal detail to evolution. However, at the time PEI did not include any

---

316 Canseco, 2020
317 Canseco, 2020
318 Wiles, 2006
319 Wiles, 2006

theories relating to evolution in the high school curriculum on the basis that it was "too controversial".[320]

Research has shown that, for both instructors and students, discussing evolution can be stressful, intimidating, and difficult to grasp conceptually.[321] Firstly, for educators, teaching evolution comes with a new set of personal challenges. Their own education on the topic could have been impacted by limited coverage in the curriculum they studied or by the educational experience they had as a youth. Each educator will have their own beliefs, preconceptions, and biases which must be managed when teaching a controversial topic. Teachers may also differ in the amount of time they spend teaching evolution, whether it is connected to other topics in a course to create a unifying theme (which often aids in making content more understandable to the audience), and how their own beliefs impact their tone and attitude whilst teaching.

While evolution is present in high school curriculums across Canadian provinces and territories, it is often a smaller topic within elective upper-level biology courses. Having evolution taught in elective courses means that a part of the student population is excluded from learning about evolution in a scientifically sound environment such as a public school. In 2021, the UCP Alberta government released a new draft K-6 curriculum which introduced the concept of evolution in grade 5, but it was widely criticized for how it handled racial and historical injustices, in addition to being accused of plagiarism and inaccuracies.[322] Under this curriculum, students would learn that "evolution over long periods of time leads to increasing complexity of organisms".[323] Currently, evolution is only part of the content for Biology 20, the elective biology course set at the grade 11 level. Evidence in support of evolution is a subtopic in the second unit of the course, titled Ecosystems and population change. The specific outcomes associated with this subtopic are the following, as written in the curriculum for Biology 20:

---

320 Wiles, 2006
321 Smith, 2009
322 Aukerman et al., 2021
323 New LearnAlberta, n.d.

20-B2.2k discuss the significance of sexual reproduction to individual variation in populations and to the process of evolution

20-B2.3k compare Lamarckian and Darwinian explanations of evolutionary change

20-B2.4k summarize and describe lines of evidence to support the evolution of modern species from ancestral forms; i.e., the fossil record, Earth's history, biogeography, homologous and analogous structures, embryology, biochemistry

20-B2.5k explain speciation and the conditions required for this process

20-B2.6k describe modern evolutionary theories; i.e., punctuated equilibrium, gradualism.[324]

While these specific learning outcomes help to specify what students are expected to understand regarding evolution, because it is part of the Biology 20 curriculum, each teacher is responsible for creating suitable tests for students. Biology 30, the grade 12 course that builds on Biology 20, has a final diploma examination which is standard across the province. This ensures that teachers cover all of the content and adequately prepare their students to encounter it on the diploma examination. The lack of a standardized exam for Biology 20 could increase the differences in how the material is covered in different classrooms. Furthermore, both Biology 20 and 30 are elective sciences courses, with Albertan high school students being able to choose which upper-level sciences courses to take, resulting in the potential lack of instruction on the theory of evolution.

Evolution is also covered in grade 11 university biology (SBI3U) in Ontario. The curriculum explains that students will explore the theory and conduct investigations related to the set topics for the course, including evolution.[325] The essential elements of Darwinism and Mendelian

---

324 Program of Study, n.d.
325 Science and Technology, n.d.

genetics are included in this course to provide students with foundational knowledge. Evolution is not included in the grade 12 biology course, grade 12 university-prep biology (SB14U), or in the curriculum for grades 1-8 that was last revised in 2007.[326] In Ontario, there is no provincial-wide testing for either high school biology course unlike in Alberta.[327] This allows for further differences to develop in how the content is taught and potentially the amount of attention given to more "controversial" topics such as evolution.

---

326 Science and Technology, n.d.
327 Tang, 2021

11

Furture Directions of Genetics & Unanswered Questions

Written By: Navneet Kang

Introduction

Despite the progress that has been made in the field of genetics, there is always more to learn. Just six decades ago, the world was busy trying to figure out how genetics worked, including the double-helical structure of DNA.[328] There has been a vast amount of knowledge gained over these past few decades and along with the field's acceleration in recent years, we are setting up a foundation needed for clinical progress. The use of technology to edit genes, such as CRISPR, provide a lot of interesting possibilities but there are ethics that must be considered. As this is a relatively new field, there is still a lot to research and uncover. Researchers all over the world are testing this new technology on animal models and beginning to optimize it so that one day it can be safe for human use as well. These optimizations can include better delivery systems and faster processes. Precision medicine is also a growing field, providing a chance for prevention, diagnosis and treatment of various conditions.[329] This field provides alternative treatment options to those patients that may

---

328 Dunn, 1991
329 Dunn, 1991

not be responding to typical treatments, such as chemotherapy. However, there are important ethical concerns, especially related to accessibility as not everyone has equal access to precision medicine. Genetic screening is becoming more popular and may eventually be used to its full potential. This field is growing exponentially and its use in clinical applications may completely change how certain diseases are treated.

CRISPR

One of the most important technologies related to genetics is Clustered Regularly Interspaced Short Palindromic Repeats (CRISPR), which allows us to edit genes.[330] These repetitive DNA sequences were initially observed in bacteria. These bacteria contained 'spacers' of DNA sequence between the repeats that exactly matched the viral sequences. It was then discovered that bacteria transcribe these DNA sequences into RNA upon viral infection. The RNA provides direction of a nuclease, which is a protein that cleaves DNA, to the viral DNA. This allows the nuclease to 'cut' the DNA, providing protection against the virus.[331] These nucleases are named "Cas," which stands for "CRISPR-associated."

With further research, it was found that RNAs could be constructed to guide a Cas nuclease towards any chosen DNA sequence.[332] This RNA guide could also be made so that it would be specific for only one sequence, ensuring that DNA will be cut only at the desired site with minimal off-target effects.[333] Further testing revealed that this technology could work in all types of cells, including human cells. Despite being extremely powerful, there are some important limitations of this technology that must be considered. It is not easy to deliver the CRISPR material in mature cells in large numbers and this hinders its clinical applications.[334] This required the use of viral vectors as a delivery method. Another limitation is that it is not 100% efficient, meaning some cells may take in CRISPR but may not show any genome editing activ-

---

330 Dance, 2015
331 Dance, 2015
332 Dance, 2015
333 Dance, 2015
334 Uddin et al., 2020

ity.[335] One other major limitation is that the results of this technology are not 100% accurate and may have off-target effects. These off-target effects are quite rare but they can result in severe consequences, particularly when used in clinical application.

This technology has profoundly changed biomedical research as it can reduce the time and expense of developing animal models with specific genomic changes.[336] This is now being used for mice specifically. It is being considered to be used in human applications as well. For example, cystic fibrosis is a human disease that has a known mutation and it is theoretically possible to insert DNA that can correct the mutation.[337] However, this has not been done yet but there are some clinical applications that are currently in human trials. One of these applications includes the engineering of T cells, which are a type of immune cells, outside of the body for cancer therapy along with editing retinal cells for an inherited form of blindness, known as leber's congenital amaurosis.[338]

To overcome the limitations of the CRISPR system, some researchers have swapped the currently used enzyme, Cas9, for Cas12a.[339] This plasmid has allowed scientists to simultaneously edit genes in 12 target sites.[340] This ensures that many genes are edited in a short period of time along with the possibility of upregulating the expression of some genes while downregulating the expression of other genes simultaneously. To add on, one of the biggest limitations has been related to delivery but researchers have created biodegradable synthetic lipid nanoparticles to deliver CRISPR into the cells.[341] This delivery system results in 90% efficacy, as all contents in the lipid nanoparticle are safely delivered to the cell.[342]

---

335 Uddin et al., 2020
336 Wilson et al., 2018
337 Wilson et al., 2018
338 Wilson et al., 2018
339 Paul & Montoya, 2020
340 Paul & Montoya, 2020
341 Wei et al., 2020
342 Wei et al., 2020

This nanoparticle, composed of disulfide bonds in the fatty chain, breaks open when inside a cell as the disulfide bonds disassemble. Continuing on, this synthetic lipid has minimal toxicity in the body and there is no limit in terms of the cargo size it can carry.[343] There is also no concern about the immune response, which is common against viral vectors. This is a major achievement as other delivery methods such as viruses and polymers have limited efficiency.[344]

Although CRISPR is often considered for editing human cells, it has the potential to completely change the agricultural field. Modern agriculture exemplifies how recent technology and research can come together to improve both crop yield and quality.[345] Despite all the advances in the agricultural field, the current methods may not be able to meet the increasing demand for food, especially considering the glocal environmental challenges that we may face in the near future. CRISPR provides a new breeding method that can produce results similar to the current methods, but they would be faster, cheaper and often more predictable.[346] Further research is needed, such as effective delivery of the CRISPR system into the right plant cells and subsequent regeneration of viable plants, before this technology can be effectively applied in the agricultural field.[347]

As explained, CRISPR allows for the editing of somatic cells, which are cells that make up most of the body, but it is also possible to edit the genomes of gametes, which are cells that make of eggs and sperm, along with early embryos.[348] The latter type of genome editing is known as germline editing.[349] When this form of editing is used in humans, it does not only impact the individual whose genome has been edited but it also impacts his or her progeny. This does create the potential to enhance desirable traits in future generations rather than curing diseases. This

---

343 Wei et al., 2020
344 Wei et al., 2020
345 Gao, 2018
346 Gao, 2018
347 Gao, 2018
348 Brokowski & Adil, 2019
349 Brokowski & Adil, 2019

situation has resulted in a moratorium being called on human germline editing, which states that although there is no permanent ban, nations will voluntarily commit to not approving any use of clinical germline editing until an international framework has been established.[350]

CRISPR is a highly useful tool and can potentially be used to treat complex diseases but it must be refined further as a technique before it can be fully used in a clinical setting. This encourages researchers to strive for improvements in this area, making the process more effective and safer for human use. Once ethical guidelines have been established as well, CRISPR will be an effective way to treat genetic diseases.

Preventative Precision Medicine

Another growing field related to genetics is precision medicine. Precision medicine, also known as personalized medicine, uses information about a person to prevent, diagnose and treat diseases using a variety of diagnostic tests.[351] This allows for the treatment to take individual variation into account, such as lifestyle, environment, microbiome and even genes.[352] This is quite important as a substantial portion of variability in drug response is genetically determined but this is not something that is often considered in the healthcare field.[353] Many other factors play a role as well, such as age, nutrition, health status, epigenetic factors and concurrent therapy, but genetics is an important factor that must be considered.

One way to apply precision medicine is for the prevention of certain diseases. This is done by creating methods that allow for the identification of those that are at risk before the disease is able to strike along with analytical tools that are able to predict which prevention strategies will work best for which patients.[354] This includes screening methods, which allow for the early identification of signs of disease, even more symptoms emerge. One example of preventative precision medicine is

---

350 Brokowski & Adil, 2019
351 Collins & Varmus, 2015
352 Collins & Varmus, 2015
353 Collins & Varmus, 2015
354 Collins & Varmus, 2015

BRCA screening. BRCA1 (BReast CAncer gene 1) and BRCA2 (BReast CAncer gene 2) are genes that are used to produce proteins responsible for repairing damaged DNA.[355] These genes are often known as tumor suppressor genes because when they have certina changes, such as mutations, it can result in cancer development.[356]

Everyone has two copies of each gene, with one copy being inherited from each parent. However, if individuals inhibit harmful variants of these genes, they have an increased risk of cancer, specifically breast and ovarian cancer.[357] Individuals with harmful variants in BRCA1 and BRCA2 also tend to develop cancer at a younger age compared to individuals who do not have the harmful variant.[358] In the general population, about 13% of women will develop breast cancer throughout their lifetime but women with a harmful BRCA1 variant have a 55%-72% chance of developing breast cancer while women with a harmful BRCA2 variant have a 45%-69% chance.[359] This demonstrates the significant role these genes play in the risk of developing breast cancer.

There are current tests available that allow us to see if someone has inherited a harmful variant of either gene.[360] This testing is typically done using a blood sample or a saliva sample. This testing is not used for the general public and is instead recommended for individuals who have a higher likelihood of carrying the harmful variants.[361] This includes individuals who have a family history of breast and ovarian cancer. Testing can be used for individuals who have not been diagnosed with cancer along with those that have been diagnosed. For those that have not been diagnosed, individuals may have the option to perform a double mastectomy along with other risk-reducing surgeries to prevent the development of cancer.[362] For those that have been diagnosed with cancer, this

---

355 Couch et al., 2014
356 Tutt & Ashworth, 2002
357 Tutt & Ashworth, 2002
358 Tutt & Ashworth, 2002
359 Tutt & Ashworth, 2002
360 Couch et al., 2014
361 Couch et al., 2014
362 Couch et al., 2014

can be important in selecting a treatment.[363]

Although this is an excellent option and can result in early detection of cancers, it is simply not an option for everyone.[364] This testing may be covered by insurance if certain criteria are met, but if these criteria are not met, it can cost anywhere from $475 to $4000.[365] If the individual is found to be carrying the harmful variant, it leads to annual cancer screenings along with the possibility of preventive surgery, which can be expensive to some as well. This makes this option practically impossible for certain individuals, which is not ideal.[366] It is important to empower individuals to get tested, which is done by developing easy access to testing methods.

Our current health system is most focused on treatment, but it is important to consider prevention, as it can result in better overall health of the population and reduced costs for the system.[367] BRCA screening is one application of preventative precision medicine, amongst many other possibilities. However, more research is needed to determine the genes responsible for certain diseases.[368] Once these relationships are better established, this will become a more viable option for many other diseases as well.

Treatment with Precision Medicine

Another branch to precision medicine involves the diagnosis and treatment. As explained earlier, there are certain genes that can play a role in the development of breast cancer, allowing for breast cancer subtypes to be created based on genetics.[369] Cancer is typically considered for precision medicine as the cause of cancer is undeniably tied to genetics.[370] Rather than receiving the standardized treatment, such as che-

---

363 Couch et al., 2014
364 Sayani, 2019
365 Sayani, 2019
366 Sayani, 2019
367 Sayani, 2019
368 Collins & Varmus, 2015
369 Collins & Varmus, 2015
370 Hodson, 2016

motherapy, researchers understand that each individual has unique genetic change and this is why treatment should be based on the genetic changes in each individual patient.[371]

One example of precision medicine being applied for treatment involves Herceptin. As mentioned earlier, breast cancer can be divided into subtypes based on genetics.[372] One subtype is known as the HER2 positive subtype, which stands for the Human Epidermal Growth Factor Receptor 2 subtype.[373] This is a member of the tyrosine kinase receptor family. In healthy individuals, approximately 20 000 HER2 receptors are expressed on healthy breast cells but in about 25% of primary breast cancers, the HER2 protein is over expressed.[374] This results in tumour cells that can have up to 2,00,000 HER2 receptors on their surface, causing uncontrollable proliferation of tumor cells.[375]

HER2 positive cancers are treated using Herceptin, a monoclonal antibody, designed especially to target HER2 receptors.[376] In simple terms, Herceptin attaches to extracellular HER2 receptors and this flags them for destruction by the immune system. It also works by preventing intracellular HER2 signaling, which inhibits cell proliferation.[377] Herceptin is also known to enhance the effects of chemotherapy based on clinical trials.[378] This means that women screened for HER2 will selectively receive Herceptin treatment rather than other standardized treatments, like chemotherapy.[379]

This field has a lot of potential, as researchers work towards treatments that will be tailored to the genetic changes in each person's cancer.[380]

---

371 Hodson, 2016
372 Hodson, 2016
373 Dean-Colomb & Esteva, 2008
374 Dean-Colomb & Esteva, 2008
375 Dean-Colomb & Esteva, 2008
376 Dean-Colomb & Esteva, 2008
377 Dean-Colomb & Esteva, 2008
378 Baselga, 2001
379 Baselga, 2001
380 Hodson, 2016

This will allow genetic tests to determine the best treatment for each patient, ensuring that the tumour responds while sparing the patient from undergoing harsh treatments that are not likely to help.[381] Many of these personalized treatments are known as targeted treatments but they are not available to everyone. The individual must be tested to see if the genetic change being targeted by the treatment is present and if it is, then the individual may be suitable for targeted therapy.[382]

This field is rapidly expanding with many ongoing clinical trials but provides a sense of hope to patients that do not respond to typical cancer treatments. However, this field is powered by patient data, which requires the genetics codes of patients along with health volunteers.[383] Collaboration between researchers, patients and volunteers is necessary for the advancement of the field.

Overall, the field of genetics has grown tremendously in the past few years and it seems that it will continue to grow. There are many powerful technologies powered by genetics, such as CRISPR and precision medicine.[384] These technologies offer treatment to individuals offering life-threatening conditions, greatly improving their quality of life. However, there is refinement needed to ensure that these options are safe, effective and accessible to the general population.[385] With further research, technologies based on genetics have the potential to transform the biomedical field.

---

381 Hodson, 2016
382 Hodson, 2016
383 Hodson, 2016
384 Uddin et al., 2020
385 Uddin et al., 2020

References

Chapter 1

Alberts, B. (2015). Molecular biology of the cell (Sixth edition). Garland Science, Taylor and Francis Group.

Allele. (n.d.). Scitable; Nature Education. Retrieved May 15, 2021, from https://www.nature.com/scitable/definition/allele-48/

Cooper, G. M. (2000). Heredity, Genes, and DNA. The Cell: A Molecular Approach. 2nd Edition. https://www.ncbi.nlm.nih.gov/books/NBK9944/

de Meeûs, T., Prugnolle, F., & Agnew, P. (2007). Asexual reproduction: Genetics and Evolutionary Aspects. Cellular and Molecular Life Sciences, 64(11), 1355–1372. https://doi.org/10.1007/s00018-007-6515-2

de Vries, J., & Archibald, J. M. (2018). Plastid genomes. Current Biology, 28(8), R336–R337. https://doi.org/10.1016/j.cub.2018.01.027

Genotype . (n.d.). Scitable; Nature education. Retrieved May 15, 2021, from https://www.nature.com/scitable/definition/genotype-234/

Kim, D. H., Yeo, S. H., Park, J.-M., Choi, J. Y., Lee, T.-H., Park, S. Y., Ock, M. S., Eo, J., Kim, H.-S., & Cha, H.-J. (2014). Genetic markers for diagnosis and pathogenesis of Alzheimer's disease. Gene, 545(2), 185–193. https://doi.org/10.1016/j.gene.2014.05.031

Koornneef, M., & Meinke, D. (2010). The Development of Arabidopsis as a Model Plant. The Plant Journal: For Cell and Molecular Biology, 61(6), 909–921. https://doi.org/10.1111/j.1365-313X.2009.04086.x

Pearson, H. (2006). What is a gene? Nature, 441(7092), 398–401. https://doi.org/10.1038/441398a

Phenotype / phenotypes . (n.d.). Scitable; Nature Education. Retrieved May 15, 2021, from https://www.nature.com/scitable/definition/phenotype-phenotypes-35/

The Human Genome Project. (n.d.). Genome.Gov. Retrieved May 14, 2021, from https://www.genome.gov/human-genome-project

Wright, S. I., Kalisz, S., & Slotte, T. (2013). Evolutionary Consequences of Self-fertilization in Plants. Proceedings of the Royal Society B: Biological Sciences, 280(1760), 20130133. https://doi.org/10.1098/rspb.2013.0133

Chapter 2:
Than, K. (2018, February 27). What is Darwin's theory of evolution? Retrieved May 15, 2021, from https://www.livescience.com/474-controversy-evolution-works.html

Sloan, P. (2019). Darwin: From Origin of Species to Descent of Man. In Stanford Encyclopedia of Philosophy. Stanford, California: Metaphysics Research Lab.

Natural selection. (2019, October 24). Retrieved May 15, 2021, from https://www.nationalgeographic.org/encyclopedia/natural-selection/

Macnow, A. S. (2020). MCAT behavioral sciences review 2021-2022. New York: Kaplan Publishing.

Birkhead, M., & Birkhead, T. (1990). The Survival Factor. New York, New York: Facts on File.

Breed, M. & Sanchez, L. (2010) Both Environment and Genetic Makeup Influence Behavior. Nature Education Knowledge 3(10):68

Sheehan, M. J., Miller, C. H., Vogt, C. C., & Ligon, R. A. (2018). Behavioral Evolution: Can You Dig It?. Current biology : CB, 28(1), R19–

R21. https://doi.org/10.1016/j.cub.2017.11.016

Gallagher, J. (2013, December 01). 'Memories' pass between generations. Retrieved May 15, 2021, from https://www.bbc.com/news/health-25156510

Szala, A., & Shackelford, T. K. (2019). Inclusive fitness. Encyclopedia of Animal Cognition and Behavior, 1-2. doi:10.1007/978-3-319-47829-6_1997-1

Gregory, T. R. (2009). Artificial selection and domestication: Modern lessons from Darwin's enduring analogy. Evolution: Education and Outreach, 2(1), 5-27. doi:10.1007/s12052-008-0114-z

Chapter 3:

Britannica, T. Editors of Encyclopaedia (2017, November 5). Colour blindness. Encyclopedia Britannica. https://www.britannica.com/science/color-blindness

Britannica, T. Editors of Encyclopaedia (2019, September 4). Albinism. Encyclopedia Britannica. https://www.britannica.com/science/albinism

Peter S. Harper. (2008). A Short History of Medical Genetics. Oxford University Press. Winchester, A. (2020, May 15). Genetics. Encyclopedia Britannica.

https://www.britannica.com/science/genetics

Chapter 4:

Khan Academy. (2021). Mendel and his peas (article). Khan Academy. https://www.khanacademy.org/science/high-school-biology/hs-classical-genetics/hs- introduction-to-heredity/a/mendel-and-his-peas.

Nature Publishing Group. (2021). Nature News: Gregor mendel, a private scientist. Nature News. https://www.nature.com/scitable/topicpage/gregor-mendel-a-private-scientist-6618227/.

Olby, R. (2021, January 15). Gregor Mendel. Encyclopedia Britannica. https://www.britannica.com/biography/Gregor-Mendel

Scoville, H. (2019). Gregor Mendel's Unique Experiments Made Him the Father of Genetics. ThoughtCo. https://www.thoughtco.com/about-gregor-mendel-1224841.

Chapter 5:
Ashraf, M. A., & Sarfraz, M. (2016). Biology and evolution of life science. Saudi journal of biological sciences, 23(1), S1–S5. https://doi.org/10.1016/j.sjbs.2015.11.012

Charlesworth, B., & Charlesworth, D. (2009). Darwin and genetics. Genetics, 183(3), 757–766. https://doi.org/10.1534/genetics.109.109991
Winther, R. (2000). Darwin on Variation and Heredity. Journal of the History of Biology, 33(3), 425-455. Retrieved May 15, 2021, from http://www.jstor.org/stable/4331610

Smýkal, P., K. Varshney, R., K. Singh, V., Coyne, C. J., Domoney, C., Kejnovský, E., & Warkentin, T. (2016). From Mendel's discovery on pea to today's plant genetics and breeding. Theoretical and Applied Genetics, 129(12), 2267–2280. https://doi.org/10.1007/s00122-016-2803-2

Talseth-Palmer, B. A., & Scott, R. J. (2011). Genetic variation and its role in malignancy. International journal of biomedical science : IJBS, 7(3), 158–171 \.

Genetic Alliance, & The New York-Mid-Atlantic Consortium for Genetic and Newborn Screening Services. (2009). Understanding Genetics: A New York, Mid-Atlantic Guide for Patients and Health Professionals. Genetic Alliance

Cooper, G. M. (2000). The Cell: A Molecular Approach. 2nd edition. Sinauer Associates.

Crummy, P. (1942). THE INCREASING IMPORTANCE OF GENETICS IN THE PREMEDICAL CURRICULUM. Proceedings of the Pennsylvania Academy of Science, 16, 106-108. Retrieved May 16,

2021, from http://www.jstor.org/stable/44109208

Charlesworth B. (2015). What Use Is Population Genetics?. Genetics, 200(3), 667–669. https://doi.org/10.1534/genetics.115.178426

Chen H. (2015). Population genetic studies in the genomic sequencing era. Dong wu xue yan jiu= Zoological research, 36(4), 223–232. https://doi.org/10.13918/j.issn.2095-8137.2015.4.223

Griffiths. (1999). Modern Genetic Analysis. New York, NY: W.H. Freeman.

Griffiths, A. (2000). An introduction to genetic analysis (7th ed.). New York, NY: W.H. Freeman.

Chapter 6:
Athira, A. (n.d.). Application of population genetics. SlideShare. https://www.slideshare.net/AthiraJanardhanan/application-of-polpulation-genetics.

Buckley, G. (2021, January 15). Allele Frequency - Definition, Calculation, Example. Biology Dictionary. https://biologydictionary.net/allele-frequency/.

Clark, A. G. (2003, December 6). Population Genetics. Encyclopedia of Genetics. https://www.sciencedirect.com/science/article/pii/B0122270800010168.

DNA Mutation and Repair. (n.d.). http://www2.csudh.edu/nsturm/CHEMXL153/DNAMutationRepair.htm#:~:text=There%20are%20three%20types%20of,base%20substitutions%2C%20deletions%20and%20inse rtions.&text=Single%20base%20substitutions%20are%20called,and%20there%20are%2 0two%20types.

Encyclopædia Britannica, inc. (n.d.). Gene flow. Encyclopædia Britannica. https://www.britannica.com/science/gene-flow#:~:text=Gene%20flow%2C%20also%20c alled%20gene,pool%20of%20the%20receiving%20population.

Encyclopædia Britannica, inc. (n.d.). J.B.S. Haldane. Encyclopædia Britannica. https://www.britannica.com/biography/J-B-S-Haldane.

Encyclopædia Britannica, inc. (n.d.). Sewall Wright. Encyclopædia Britannica. https://www.britannica.com/biography/Sewall-Wright.

Encyclopædia Britannica, inc. (n.d.). Sir Ronald Aylmer Fisher. Encyclopædia Britannica. https://www.britannica.com/biography/Ronald-Aylmer-Fisher.

Khan Academy. (n.d.). Allele frequency & the gene pool (article). Khan Academy. https://www.khanacademy.org/science/ap-biology/natural-selection/hardy-weinberg-equil ibrium/a/allele-frequency-the-gene-pool.

Khan Academy. (n.d.). Natural selection in populations (article). Khan Academy. https://www.khanacademy.org/science/ap-biology/natural-selection/population-genetics/a/natural-selection-in-populations.

Khan Academy. (n.d.). Population genetics: When Darwin met Mendel (video). Khan Academy. https://www.khanacademy.org/science/biology/crash-course-bio-ecology/crash-course-bi ology-science/v/crash-course-biology-117.

Millstein, R. L., & Skipper, R. A. (2007, October). Population Genetics. Google Books. https://books.google.ca/books?id=aZOgg-x4UyIC&pg=PA24&lpg=PA24&dq=Evolution%2Bis%2Bdriven%2Bprimarily%2Bby%2Bnatural%2Bselection%2C%2Bor%2Bmass%2 2Bselection%2C%2Bat%2Blow%2Blevels%2Bacting%2Bon%2Bthe%2Baverage%2Bef fects%2Bof%2Bsingle%2Ballele%2Bchanges%2B%28of%2Bweak%2Beffect%29%2Bat%2Bsingle%2Bloci%2Bindependent%2Bof%2Ball%2Bother%2Bloci.&source=bl&ots=_nb50FAuaw&sig=ACfU3U0dWFz-a14TYZDB-J6Rc7nnp_7xu7w&hl=en&sa=X&ved= 2ahUKEwiSqoKA0MzwAhVGU98KHYuKA-AQ6AEwCnoECA0QAw#v=onepage&q& f=false.

Okasha, S. (2012, July 5). Population Genetics. Stanford Encyclopedia of Philosophy. https://plato.stanford.edu/entries/population-genetics/#HarWeiPri.

Population Genetics of Plant Pathogens Genetic Drift. The American Phytopathological Society (APS). (n.d.). https://www.apsnet.org/edcenter/disimpactmngmnt/topc/PopGenetics/Pages/GeneticDrift. aspx#:~:text=Genetic%20drift%20is%20a%20process,a%20short%20 period%20of%20ti me.

Population genetics. University of Leicester. (2017, August 1). https://www2.le.ac.uk/projects/vgec/highereducation/topics/population-genetics#:~:text= Factors%20influencing%20the%20genetic%20diversity,and%20non%2Drandom%20mat ing%20patterns.

ScienceDaily. (n.d.). Allele. ScienceDaily. https://www.sciencedaily.com/terms/allele.htm.

SparkNotes. (n.d.). SparkNotes. https://www.sparknotes.com/biology/evolution/populationgenetics/summary/.

What is evolution? yourgenome. (2017, February 17). https://www.yourgenome.org/facts/what-is-evolution#:~:text=In%20biology%2C%20evolution%20is%20the,and%20gradually%20change%20over%20time.

Chapter 7:

About Cri du Chat Syndrome. Genome.gov. (n.d.). https://www.genome.gov/Genetic-Disorders/Cri-du-Chat.

Clancy, S. (2008). Nature News. https://www.nature.com/scitable/topicpage/genetic-mutation-441/.

Centers for Disease Control and Prevention. (2020, October 23). Data and Statistics on Down Syndrome. Centers for Disease Control and Prevention. https://www.cdc.gov/ncbddd/birthdefects/downsyndrome/data.html.

Collins, F. S. (n.d.). Copy Number Variation (CNV). Genome.gov.

https://www.genome.gov/genetics-glossary/Copy-Number-Variation.

Loewe, L. (2008). Genetic Mutation. Nature News. https://www.nature.com/scitable/topicpage/genetic-mutation-1127/.

Morris, J. (2019). Biology: how life works. W.H. Freeman & Company.

Myers R. H. (2004). Huntington's disease genetics. NeuroRx : the journal of the American Society for Experimental NeuroTherapeutics, 1(2), 255–262. https://doi.org/10.1602/neurorx.1.2.255

Resta, R. (2016, July 21). Miracles, Monsters, And Do-Re-Mi: A Variant Cultural History of The Word "Mutation". The DNA Exchange. https://thednaexchange.com/2016/07/20/miracles-monsters-and-do-re-mi-a-variant-cultur al-history-of-the-word-mutation/.

U.S. Department of Health and Human Services. (n.d.). FG syndrome. Genetic and Rare Diseases Information Center. https://rarediseases.info.nih.gov/diseases/2317/fg-syndrome.

Yagi T. (2017). A perspective of Genes and Environment for the development of environmental mutagen research in Asia. Genes and environment : the official journal of the Japanese Environmental Mutagen Society, 39, 23. https://doi.org/10.1186/s41021-017-0083-y

Zehr, E. P. (2014, June 5). We're All X-Men as Far as Genetic Mutations Go. Scientific American Blog Network. https://blogs.scientificamerican.com/guest-blog/we-re-all-x-men-as-far-as-genetic-mutati ons-go/.

Chapter 8:
Robinson, J. L., PhD. (2019). Animal cloning. Salem Press Encyclopedia of Health.

Rugnetta, M. (2020, April 15). Cloning. Encyclopedia Britannica. https://www.britannica.com/science/cloning

National Geographic Society. (2019, July 8).Cloning. National Geographic. https://www.nationalgeographic.org/encyclopedia/cloning/

Buratovich, M. A. . B. S. . M. A. . P. D. (2018). Science of cloning. Salem Press Encyclopedia of Science.

Uddin, F., Rudin, C. M., & Sen, T. (2020). CRISPR Gene Therapy: Applications, Limitations, and Implications for the Future. Frontiers in Oncology, 10.

Watts, C. P. D. . B. A. S. . B. S. (2020). Genetically Modified Organisms. Salem Press Encyclopedia of Science.

Weasel, Lisa H. Food Fray: Inside the Controversy over Genetically Modified Food. New York: Amacom-American, 2009. Print.

Thieman, William J., and Michael A. Palladino. (2013) Introduction to Biotechnology. 3rd ed.

Boston: Pearson. Print.

Boylan, Michael. (2000). Genetic Engineering. In Medical Ethics, edited by Boylan. Upper Saddle River, N.J.: Prentice Hall. Print.

Cibelli, Jose B., Robert P. Lanza, Michael D. West, and Carol Ezzell. (2002). The First Human Cloned Embryo. Scientific American. Vol. 286, no. 1. 44–48. Print.

Owen, H. R. (2019). Cloning of plants. Salem Press Encyclopedia of Science. Houdebine, Louis-Marie. (2003) Animal Transgenesis and Cloning. Trans. Louis-Marie

Houdebine et al. Hoboken: Wiley. Print.

Jabr, Ferris. (2013). Will Cloning Ever Save Endangered Animals? Scientific American. Nature America. Web.

Chapter 9:

About Sickle Cell Disease. (2020, May 26). National Human Genome Research Institute. https://www.genome.gov/Genetic-Disorders/Sickle-Cell-Disease#

Anzilotti, A. W., MD. (2021, March). Gene Changes (Mutations). KidsHealth from Nemours. https://kidshealth.org/en/parents/gene-mutations.html#

Armstrong-Carter, E., Wertz, J., & Domingue, B. W. (2021). Genetics and Child Development: Recent Advances and Their Implications for Developmental Research. Child Development Perspectives, 15(1), 57–64. https://srcd.onlinelibrary.wiley.com/doi/epdf/10.1111/cdep.12400 Asendorpf, J. B. (2001). Temperament: Familial Analysis and Genetic Aspects. International Encyclopedia of the Social and Behavioral Sciences. Published. https://www.sciencedirect.com/topics/computer-science/heritability-estimate

Baggini, J. (2015, March 19). Do your genes determine your entire life? The Guardian. https://www.theguardian.com/science/2015/mar/19/do-your-genes-determine-your-entire- life

BRCA: The Breast Cancer Gene. (n.d.). National Breast Cancer Foundation, Inc. Retrieved May 16, 2021, from https://www.nationalbreastcancer.org/what-is-brca#

Breed, M., & Sanchez, L. (2010). Both Environment and Genetic Make-up Influence Behavior.

Nature Education Knowledge, 3(10), 68. https://www.nature.com/scitable/knowledge/library/both-environment-and-genetic-makeu   p-influence-behavior-13907840/#

Cherry, K., & Block D.B. (2020, March 16). How Genes Influence Child Development. Verywell Mind. https://www.verywellmind.com/genes-and-development-2795114#

Collins, F. S., MD, PhD. (n.d.). Mutation. National Human Genome Research Institute. Retrieved May 16, 2021, from https://www.genome.

gov/genetics-glossary/Mutation#

Department of Health, Government of Western Australia. (n.d.). Genetic conditions. Health Information for Western Australians. Retrieved May 16, 2021, from https://healthywa.wa.gov.au/Articles/F_I/Genetic-conditions

DNA DETERMINES YOUR APPEARANCE! (n.d.). PennState Center for Nanoscale Science.
Retrieved May 16, 2021, from https://www.mrsec.psu.edu/content/dna-determines-your-appearance#

Genetics and Behavior. (n.d.). Lumen. Retrieved May 16, 2021, from https://courses.lumenlearning.com/boundless-psychology/chapter/genetics-and-behavior/

Grisham, J. (2019, March 27). Some People Who Need a Bone Marrow Transplant Will Never Find a Donor — and What Can Be Done about It. Memorial Sloan Kettering Cancer Center. https://www.mskcc.org/news/some-people-need-bone-marrow-transplant-never-find-donor-what-can-done-about-it

Hernandez, L. M., & Blazer, D. G. (Eds.). (2006). Genes, Behavior, and the Social Environment: Moving Beyond the Nature/Nurture Debate. National Academies Press (US). Published. https://www.ncbi.nlm.nih.gov/books/NBK19932/

How do genes impact health and disease? (n.d.). Genes in Life. Retrieved May 16, 2021, from http://www.genesinlife.org/genes-your-health/how-do-genes-impact-health-and-disease

How Does Genetics Affect Our Lives? (n.d.). BYJU'S. Retrieved May 16, 2021, from https://byjus.com/questions/how-does-genetics-affect-our-lives/

How Genetics Impacts Your Everyday Life and Wellness. (n.d.). Swagene. Retrieved May 16, 2021, from http://swagene.com/news/how-genetics-impacts-your-everyday-life-introducing-swaviva/

How to Prevent Heart Disease. (n.d.). MedlinePlus. Retrieved May 16, 2021, from https://medlineplus.gov/howtopreventheartdisease.html

Is eye color determined by genetics? (2021, May 12). MedlinePlus. https://medlineplus.gov/genetics/understanding/traits/eyecolor/#

Is hair color determined by genetics? (2020, September 17). MedlinePlus. https://medlineplus.gov/genetics/understanding/traits/haircolor/#

Khan Academy. (n.d.). Genes, environment, and behavior. Retrieved May 16, 2021, from https://www.khanacademy.org/test-prep/mcat/behavior/behavior-and-genetics/a/genes-en vironment-and-behavior

Kramer, C., & Patel, H. (2010). Genetics in Everyday Life. GENIE, University of Leicester. https://www2.le.ac.uk/projects/vgec/downloads/basics/genetics-in-everyday-life-transcrip ts/Script%20for%20Genetics%20in%20Everyday%20life%20vid.docx#

Kreuger, R. F., South, S., Johnson, W., & Lacono, W. (2008). The Heritability of Personality is not Always 50%: Gene-Environment Interactions and Correlations between Personality and Parenting. Journal of Personality, 76(6), 1485–1522. https://www.ncbi.nlm.nih.gov/pmc/articles/PMC2593100/

Learned Behaviors In Humans. (2019, September 16). Cabarrus. https://www.cabarrus.k12.nc.us/site/handlers/filedownload.ashx?-moduleinstanceid=1389 50&dataid=391181&FileName=Learned%20Behaviors.pdf

Mayo Clinic Staff. (2017, November 18). Turner syndrome. Mayo Clinic. https://www.mayoclinic.org/diseases-conditions/turner-syndrome/symptoms-causes/syc-2 0360782#

Mayo Clinic Staff. (2018, June 23). Cataracts. Mayo Clinic. https://www.mayoclinic.org/diseases-conditions/cataracts/symptoms-causes/syc-2035379 0#

Mayo Clinic Staff. (2019a, September 4). Medical history: Compiling your medical family tree.

Mayo Clinic. https://www.mayoclinic.org/healthy-lifestyle/adult-health/in-depth/medical-history/art-20 044961#

Mayo Clinic Staff. (2019b, September 21). Klinefelter syndrome. Mayo Clinic. https://www.mayoclinic.org/diseases-conditions/klinefelter-syndrome/symptoms-causes/s yc-20353949

Mehta, S. (n.d.). Ask Dr. M: How should I prepare to see a new doctor? Dot Health. Retrieved May 16, 2021, from https://www.dothealth.ca/post/ask-dr-m-how-should-i-prepare-to-see-a-new-doctor#

Oculocutaneous albinism. (2020, August 18). MedlinePlus. https://medlineplus.gov/genetics/condition/oculocutaneous-albinism/#inheritance

predisposition. (n.d.). In Cambridge Advanced Learner's Dictionary & Thesaurus. Cambridge University Press. Retrieved May 16, 2021, from https://dictionary.cambridge.org/dictionary/english/predisposition

recurrent cancer. (n.d.). In National Cancer Institute. National Cancer Institute at the National Institutes of Health. Retrieved May 16, 2021, from https://www.cancer.gov/publications/dictionaries/cancer-terms/def/recurrent-cancer

Riesen, G. (2019, February 15). Hair Color. The Tech Interactive. https://genetics.thetech.org/ask-a-geneticist/hair-color-genetics

Robertson, S., B. Sc., & Anderton, K., B. Sc. (2019, September 15). Genetics of Eye Color. News-Medical.https://www.news-medical.net/health/Genetics-of-Eye-Color.aspx

Sokol, R. (2021, January 28). Is Personality Genetic? Family Education. https://www.familyeducation.com/genetics-pregnancy/is-personality-genetic

Stages 0 & 1. (n.d.). National Breast Cancer Foundation, Inc. Retrieved May 16, 2021, from https://www.nationalbreastcancer.org/breast-cancer-stage-0-and-stage-1#

temperament. (n.d.). In Cambridge Advanced Learner's Dictionary & Thesaurus. Cambridge University Press. Retrieved May 16, 2021, from https://dictionary.cambridge.org/dictionary/english/temperament

The benefits of genetics. (n.d.). Biron. Retrieved May 16, 2021, from https://www.biron.com/en/education-center/neat-little-guide/genetics/

Thompson, R. (2021). Social and personality development in childhood. In R. Biswas-Diener &

E. Deiner (Eds.), Noba textbook series: Psychology (p. unknown). DEF Publishers. https://nobaproject.com/modules/social-and-personality-development-in-childhood#conte nt

Treatments for breast cancer. (n.d.). Canadian Cancer Society. Retrieved May 16, 2021, from https://www.cancer.ca/en/cancer-information/cancer-type/breast/treatment/?region=on

Walinga, J., & Stangor, C. (n.d.). 12.3 Is Personality More Nature or More Nurture? Behavioural and Molecular Genetics. In Introduction to Psychology (1st Canadian Edition, p. unknown). Pressbooks. https://opentextbc.ca/introductiontopsychology/chapter/11-3-is-personality-more-nature-o r-more-nurture-behavioral-and-molecular-genetics/#

What does it mean to have a genetic predisposition to a disease? (n.d.). MedlinePlus. Retrieved May 16, 2021, from https://medlineplus.gov/genetics/understanding/mutationsanddisorders/predisposition/

What is a gene variant and how do variants occur? (2021, March 25). MedlinePlus. https://medlineplus.gov/genetics/understanding/mutationsanddisorders/genemutation/

What is Down Syndrome? (2021, April 6). Centers for Disease Control and Prevention. https://www.cdc.gov/ncbddd/birthdefects/downsyndrome.html

What is Genotyping? (n.d.). ThermoFisher Scientific. Retrieved May 16, 2021, from https://www.thermofisher.com/ca/en/home/life-science/pcr/real-time-pcr/real-time-pcr-lea   rning-center/genotyping-analysis-real-time-pcr-information/what-is-genotyping.html#per

What is Sickle Cell Disease? (2020, December 14). Centers for Disease Control and Prevention. https://www.cdc.gov/ncbddd/sicklecell/facts.html#

What kinds of gene variants are possible? (n.d.). MedlinePlus. Retrieved May 16, 2021, from https://medlineplus.gov/genetics/understanding/mutationsanddisorders/possiblemutations/

Chapter 10:
Aukerman, M., Burwell, C., & Scott, D. (2021, April 7). By forgetting about thinking, Alberta's curriculum draft misses the mark. Cbc.Ca. https://www.cbc.ca/news/canada/calgary/road-ahead-alberta-education-curriculum-critici sm-1.5978023

Ayala, F. Jose (2020, November 20). Evolution. Encyclopedia Britannica. https://www.britannica.com/science/evolution-scientific-theory

Bielo, J. S. (2018). Ark Encounter: The Making of a Creationist Theme Park. NYU Press.

Canseco, B. M. (2020, February 17). Most Canadians Believe Human Beings on Earth Evolved. Research Co. https://researchco.ca/2019/12/04/creationism-evolution-canada/

Draper, P. (2002). Irreducible Complexity and Darwinian Gradualism. Faith and Philosophy,
19(1), 3–21. https://doi.org/10.5840/faithphil20021912

Knight, C. T. E. (2018, July 6). Ark Encounter reports 1 million visitors in second year. Cincinnati Enquirer. https://eu.cincinnati.com/story/news/2018/07/05/ark-encounter-reports-1-million-visitors- 2nd-year/759704002/

Science and Technology. (n.d.). Ontario Ministry of Education. http://www.edu.gov.on.ca/eng/curriculum/elementary/scientec.html

New LearnAlberta | Science. (n.d.). Learn Alberta. https://curriculum.learnalberta.ca/curriculum/en/c/sci5

New Revised Standard Version Catholic Edition Bible. (1989). Harper Catholic Bibles.

Oerter, R. (2006). Does Life On Earth Violate the Second Law of Thermodynamics. Physics and Astronomy Department George Mason University. http://physics.gmu.edu/%7Eroerter/EvolutionEntropy.htm#:%7E:text=Evolution%2C%2 0the%20argument%20goes%2C%20 is,evolution%20violates%20the%20second%20law.   &text=Rather%2C%20the%20second%20law%20says,the%20whole%20system%20mus t%20increase.

Olson, A. C. (2017, January 4). Evolution and Mass Media. WordPress. https://seemsobvioustome.wordpress.com/2017/01/04/evolution-and-mass-media/

Program of Study - LearnAlberta.ca. (n.d.). Learn Alberta. https://www.learnalberta.ca/ProgramOfStudy.aspx?lang=en&ProgramId=864871#

Series, M. H. R. (2006). What is Darwinism? By Charles Hodge . . . Scholarly Publishing Office, University of Michigan Library.

Smith, M. U. (2009). Current Status of Research in Teaching and Learning Evolution: II. Pedagogical Issues. Science & Education, 19(6–8), 539–571. https://doi.org/10.1007/s11191-009-9216-4

Steinberg, A. (2010). The Theory of Evolution - A Jewish Perspective. Rambam Maimonides Medical Journal, 1(1), e0008. https://doi.org/10.5041/rmmj.10008

Sterman, B. (1994). Judaism and Darwinian Evolution. Tradition: A Journal of Orthodox Jewish Thought, 29(1), 48-75. Retrieved May 11, 2021, from http://www.jstor.org/stable/23260873

Wiles, Jason. (2006). Evolution in Schools: Where's Canada?. Education Canada. 46. 37-41.

Wiki Targeted (Entertainment). (n.d.). X-Men Movies Wiki. https:// xmenmovies.fandom.com/wiki/Mutant_Abilities

Wiki Targeted (Entertainment). (n.d.). Simpsons Wiki. https://simpsons. fandom.com/wiki/Homer%27s_Evolution_couch_gag

Chapter 11:

Dunn, L. C. (1991). A short history of genetics: The development of some of the main lines of thought: 1864-1939. A Short History of Genetics: The Development of Some of the Main Lines of Thought: 1864-1939. https://www.cabdirect.org/cabdirect/abstract/19921630111

Dance, A. (2015). Core Concept: CRISPR gene editing. Proceedings of the National Academy of Sciences, 112(20), 6245–6246. https://doi. org/10.1073/pnas.1503840112

Uddin, F., Rudin, C. M., & Sen, T. (2020). CRISPR Gene Therapy: Applications, Limitations, and Implications for the Future. Frontiers in Oncology, 10. https://doi.org/10.3389/fonc.2020.01387

Wilson, L. O. W., O'Brien, A. R., & Bauer, D. C. (2018). The Current State and Future of CRISPR-Cas9 gRNA Design Tools. Frontiers in Pharmacology, 9. https://doi.org/10.3389/fphar.2018.00749

Paul, B., & Montoya, G. (2020). CRISPR-Cas12a: Functional overview and applications. Biomedical Journal, 43(1), 8–17. https://doi. org/10.1016/j.bj.2019.10.005

Wei, T., Cheng, Q., Min, Y.-L., Olson, E. N., & Siegwart, D. J. (2020). Systemic nanoparticle delivery of CRISPR-Cas9 ribonucleoproteins for effective tissue specific genome editing. Nature Communications, 11(1), 3232. https://doi.org/10.1038/s41467-020-17029-3

Gao, C. (2018). The future of CRISPR technologies in agriculture. Nature Reviews Molecular Cell Biology, 19(5), 275–276. https://doi.

org/10.1038/nrm.2018.2

Brokowski, C., & Adli, M. (2019). CRISPR ethics: Moral considerations for applications of a powerful tool. Journal of Molecular Biology, 431(1), 88–101. https://doi.org/10.1016/j.jmb.2018.05.044

Collins, F. S., & Varmus, H. (2015). A New Initiative on Precision Medicine. New England Journal of Medicine, 372(9), 793–795. https://doi.org/10.1056/NEJMp1500523

Couch, F. J., Nathanson, K. L., & Offit, K. (2014). Two Decades After BRCA: Setting Paradigms in Personalized Cancer Care and Prevention. Science, 343(6178), 1466–1470. https://doi.org/10.1126/science.1251827

Tutt, A., & Ashworth, A. (2002). The relationship between the roles of BRCA genes in DNA repair and cancer predisposition. Trends in Molecular Medicine, 8(12), 571–576. https://doi.org/10.1016/S1471-4914(02)02434-6

Sayani, A. (2019). Inequities in genetic testing for hereditary breast cancer: Implications for public health practice. Journal of Community Genetics, 10(1), 35–39. https://doi.org/10.1007/s12687-018-0370-8

Hodson, R. (2016). Precision medicine. Nature, 537(7619), S49–S49. https://doi.org/10.1038/537S49a

Dean-Colomb, W., & Esteva, F. J. (2008). Her2-positive breast cancer: Herceptin and beyond. European Journal of Cancer, 44(18). https://doi.org/10.1016/j.ejca.2008.09.013

Baselga, J. (2001). Herceptin® Alone or in Combination with Chemotherapy in the Treatment of HER2-Positive Metastatic Breast Cancer: Pivotal Trials. Oncology, 61(Suppl. 2), 14–21. https://doi.org/10.1159/000055397

www.ingramcontent.com/pod-product-compliance
Lightning Source LLC
Chambersburg PA
CBHW070356200326
41518CB00012B/2247